SpringerBriefs in Water Science and Technology

SpringerBriefs in Water Science and Technology present concise summaries of cutting-edge research and practical applications. The series focuses on interdisciplinary research bridging between science, engineering applications and management aspects of water. Featuring compact volumes of 50 to 125 pages (approx. 20,000–70,000 words), the series covers a wide range of content from professional to academic such as:

- Literature reviews
- In-depth case studies
- Bridges between new research results
- Snapshots of hot and/or emerging topics

Topics covered are for example the movement, distribution and quality of freshwater; water resources; the quality and pollution of water and its influence on health; and the water industry including drinking water, wastewater, and desalination services and technologies.

Both solicited and unsolicited manuscripts are considered for publication in this series.

Shakir Ali

Traditional Water Conservation Community-Managed Structures and Their Role in Valley Dwellers' Livelihoods

A Case Study of Kangra Valley, Himachal Pradesh, Western Himalayas

Shakir Ali ⓘ
Department of Civil Engineering Science
Faculty of Engineering & the Built
Environment, Auckland Park
University of Johannesburg
Johannesburg, Gauteng, South Africa

ISSN 2194-7244 ISSN 2194-7252 (electronic)
SpringerBriefs in Water Science and Technology
ISBN 978-3-032-04639-0 ISBN 978-3-032-04637-6 (eBook)
https://doi.org/10.1007/978-3-032-04637-6

This Springer imprint is published by the registered company Springer Nature Switzerland AG
The registered company address is: Gewerbestrasse 11, 6330 Cham, Switzerland

Preface

The book *Traditional Water Conservation Community-Managed Structures and Their Role in Valley Dwellers' Livelihoods: A Case Study of Kangra Valley, Himachal Pradesh, Western Himalayas* summarised several indigenous water conservation structures of the Kangra Valley of Himachal Pradesh and their role in valley dwellers livelihoods. The book explores how valley dwellers used traditional water systems for boosting their socio-economic conditions, showcasing their natural resources management wisdom. Furthermore, the book recognizes and values local communities water management practices in a challenging environment, such as Himalayas.

Kangra Valley is located in the Western Himalayas, India, and bestowed with natural wonders, such as stunning landscapes, snow covered Dhauladhar mountains, flowing perennial streams networks (Khad), springs, and many more. The valley envelops everyone and attracts numerous national and international tourists every year with its beautiful landscapes. A calm place away from cities becomes a hotspot for meditation in the hill stations of Kangra Valley and develops as a centre of meditation and has a strong Tibetan cultural influence, garnishing an additional cultural richness.

People of Kangra are friendly, quiet, hospitable and welcome everyone, and never fear to share their rich cultural heritages. A large flow of tourists in the valley also supports local people, generate employments and a major source of earning after agriculture.

This book discusses how valley dwellers used traditional knowledge for their sustainable development. I argued the importance of traditional water conservation structures in the long run and asserted their preservation in the valley. The book highlightes how community engagement is imperative for effective management of natural resources and developing long-term sustainable solutions. I also asserted on capacity building through community training about best water management practices, which is quite lacking or weakly structured and often fragmented. The book digs out how valley dwellers are using nature-based solutions for their socio-economic development and enhancing their businesses.

The book underscores the importance of farmer's wisdom in improving agricultural productivity and strengthening their livelihoods. Furthermore, the book discusses the need for revival of water systems for building the water resilience valley because water management is imperative for rural transformation and self-reliance. It also emphasize how proper institutional framework and local community's synergies are crucial in addressing resource scarcity.

The book also discusses how valley dwellers survived on their traditional water system and the role of community participation in fighting against any harsh challenges. The book further explores the raising water issues and highlights the urgent pressing challenges to the concerned institutions. It also discusses the importance of developing cutting-edge technologies to foster tailored solutions without neglecting traditional knowledge. Few invaluable insights into the pressing issues are also highlighted. Lastly, the book underscores the institution's role in addressing water and food scarcity under present climate change conditions.

The book explores traditional indigenous knowledge and its interconnectedness with the agro-economy. However, with time, numerous water conservation structures have been reduced and many of them are defunct. The shift to tap water resulted in leaving other natural water structures unmaintained and unpreserved. The traditional water structures have provided long-term sustainable benefits to the local communities for generations, which are ruining with modernization.

My hands-on experience while conducting the field survey was unforgettable. I thank farmers, elders, and numerous family members whom I interacted with in the Kangra Valley. I thank people of Kangra for their supportive behaviour and providing valuable information. And for better communication, most of the discussions were in the local Kangri/Gaddi language so that responders could better express themselves. Responders were also unaware of the book idea so that they could honestly express their views.

The book is structured in five chapters, divided into three main parts, with an additional section for field questionnaires:

- In *Part I*, I placed two chapters i.e., Chaps 1 and 2.

Chapter 1, "**Introduction to Water Conservation Traditional Systems in the Kangra Valley of Himachal Pradesh: Insights from Agro-economy,**" introduces readers to various traditional water conservation structures and their importance in agricultural productivity, where more than half of the valley dwellers are engaged in farming for their livelihoods. The chapter also gives a brief introduction to the current situation of the Kangra Valley.

Chapter 2, "**Indigenous Water Conservation Structures in the Kangra Valley of Himachal Pradesh**" discusses the importance of several traditional water systems of the valley. For instance, the chapter discusses boweris, springs, khads, khatris, etc., and their importance in enduring resilient and sustainable water future for the valley.

- In *Part II*, I kept one chapter, i.e., Chap. 3.

Chapter 3 "**Kangra Valley's Kuhl System: The Last Hope for Irrigation in Peril**" discusses the Kuhl system in the Kangra Valley. The chapter on Kuhls highlights their importance to farmers in enhancing crop productivity. A detailed discussion on Kuhl irrigation provides insight into using traditional wisdom to grow multiple crops annually, thereby doubling farmer's income. Therefore, a separate chapter is dedicated on Kuhls.

- In the last part (***Part III***), I placed two chapters i.e., Chaps. 4 and 5.

Chapter 4 "**Role of Institutions in Kangra Valley Development**" underscores the role of institutions in maintaining traditional water resources.

Finally, the last chapter, Chap. 5 "**Towards Sustainable Development in the Kangra Valley: Challenges and Prospects**" discusses the challenges and the way forward for the sustainable development of the valley and ends with recommendations and conclusions.

- Furthermore, **Questionnaires** can be found at book end.

The book is an outcome of my long, decadal, and intermittent discussions over the years. My growing interest in the water systems and cultural heritage of the Kangra Valley prompted me to examine its water conservation structures and the socio-economy of valley dwellers in detail. This makes it possible for me to compile this book. For this study, intermittent field surveys were conducted from 2014 until 2024. I here acknowledge the field support from Mr. Satish, Regional Centre Dharamshala, from 2014 to 2017. The study was conducted without any assistance from any funding agencies, and the views presented in the book are solely my own and nothing to do with my working institutions. Additionally, the opinions presented in the book are drawn from my wisdom based on the knowledge that I gathered during this vast span. However, I only conducted limited fieldworks due to many inaccessible terrains, family responsibilities, and limited personal work capital. For any further discussion, I can be contacted at shakiriitb@gmail.com.

I believe that such studies should be further continued to gain deeper insights into traditional indigenous knowledge and their importance in bringing agricultural reforms to enhance agricultural productivity for the self-reliance of the Kangra Valley dwellers. The book represents my modest effort in bringing few insights, which could be useful to researchers, students, institutions, stakeholders, and local communities.

It is duly acknowledged that I used ChatGPT-4 solely for the purpose of refining and improving the English language. All ideas, arguments, field insights, and the overall structure are entirely my own.

Last but not the least, I thank ***Margaret Deignan*** and the entire Springer team for their continuous support throughout the book preparation.

Dharamshala, India   Dr. Shakir Ali
July 2025

Competing Interests The author has no competing interests to declare that are relevant to the content of this manuscript.

Contents

Part I
An Introduction to Traditional Water Conservation Structures and Their Role in Enhancing Agricultural Productivity

Chapter 1
Introduction to Water Conservation Traditional Systems in the Kangra Valley of Himachal Pradesh: Insights from Agro-economy

Abstract As humankind evolved, water, food, and shelter became the primary necessities for survival however, traditional knowledge ensured the ease for survival. Over time, humans invented several survival skills and with wisdom, developed strategies to survive and advance their socio-economic conditions. Such practices have been well preserved in the mountainous regions of India. The chapter examines these traditional systems in the Kangra Valley, located in Himachal Pradesh, Western Himalayas. The valley is one of the largest in the state with its population, has a unique culture and known for its traditional indigenous heritages. The study employed a systematic search to collect the literature on traditional water heritages along with agro-economy. The study found that the valley has several traditional water conservation heritages, such as *Kuhls, Boweris, Khatri's*, and *dug-wells*. These structures have supported the valley dwellers for generations and enhanced their socio-economic conditions. This chapter highlights how these water conservation structures played a pivotal role for survival and were an important source of livelihood, are now facing serious environmental challenges.

Keywords Himachal Pradesh · Western Himalayas · Indigenous heritage · Kangra Valley · Water and food security

1.1 Introduction

Water, food, and a clean environment are essential for humans to live a better life. Numerous studies highlight that access to these resources is crucial for maintaining peace, harmony, and internal security (Krampe 2016; Krampe et al. 2021; Simangan et al. 2022). Indeed, water and food plays a vital role in alleviating poverty, hunger, and inequality. However, limiting access to water and food can escalate conflicts and social unrest (Unfried et al. 2022). Humans realised such insecurities and developed several innovative water conservation structures, leveraging indigenous knowledge to enhance agricultural productivity, and food security. For instance, Malapane et al. (2024) documented how community indigenous knowledge in South Africa

S. Ali, *Traditional Water Conservation Community-Managed Structures and Their Role in Valley Dwellers' Livelihoods*, SpringerBriefs in Water Science and Technology, https://doi.org/10.1007/978-3-032-04637-6_1

has sustained communities for generations. Similarly, Khaneiki (2020) documented how indigenous knowledge nourishes communities in Iran. Baker (2007) also illustrated how indigenous knowledge boosted agricultural productivity in the Western Himalayas. Furthermore, Darwish et al. (2024) highlights how traditional irrigation shows remarkable results in agriculture. Gaurav et al. (2025) also reported on unique community-led irrigation systems in the plains of North India. These case studies underscore the importance of traditional indigenous knowledge, showcasing its role in sustainable water management, advancing agriculture, and overall supporting communities.

Numerous traditional water conservation structures are found across Himalayas and are imperative for the communities' livelihoods (Rautela 2015). For instance, Baker (2007) pointed out, how man-made gravity water channels could enhance annual agricultural productivity in the Western Himachal. Agarwal et al. (2024) discussed how Himalayan communities relied on natural resources for their livelihoods. Negi et al. (2025) underscore the importance of traditional ecological knowledge in shaping climate resilient villages in the Himalayas. Thus, local knowledge is crucial for transforming farming systems and contributes significantly in designing climate-resilient agriculture (Ali et al. 2025).

In fact, traditional indigenous knowledge is deeply embedded in cultural practices and a way towards prosperity (Aliabadi et al. 2022; Hansen et al. 2024). However, traditional knowledge often clashes with modern technologies. As a result, traditional systems are often replaced by new advancements and the change is clearly evident globally. Adopting modern technologies also reflects cultural identity loss and a sense of insecurity. Furthermore, diminishing the traditional knowledge over Western technologies appears inevitable, the importance of indigenous knowledge remains significant, particularly in addressing the ongoing water and food crises amid extreme climate change. Nevertheless, some forms of indigenous knowledge have been well preserved in Himachal Pradesh and continue to thrive.

Himachal Pradesh is a northern state of India, largely surrounded by the tectonically formed young Himalayas. The state enjoys various climates—from subtropical to arid cold temperate, which vary with elevation. Himachal Pradesh is relatively a small state of India in terms of area and population (Asian Development Bank 2010). The state's economy is primarily driven by agriculture and tourism. It is home to numerous tourist destinations and often referred as the "fruit basket of India", a title earned for its abundant fruit production.

Mountains have always fascinated humans however, the unsung reality of hilly stations is, they are increasingly grappling with water scarcity.[1] There are growing water concerns in the mountains (Zaryab et al. 2025). Water scarcity directly affect agricultural productivity, and a great concern in Himachal Pradesh, where over 60% of the population is engaged (https://hpshiva.hp.gov.in/). One such region is the Kangra Valley of Himachal Pradesh, which is currently facing numerous challenges

---

[1] https://dialogue.earth/en/climate/springs-dying-across-himachal-pradesh/ (Assessed on 25–01–2025).

in addressing urgent water and food crises that are further intensified by growing climate uncertainties.

Kangra Valley is well known for its popular tourist destinations (Kumar 2023). Notable tourist destinations are Dharamshala, McLeod Ganj, Dharamkot, Bhagsu, Palampur, etc. Dharamshala is also an administrative headquarter[2] of Kangra District and has a large influence of Tibetan culture and thus also known as "Little Lhasa". It also serves as the winter capital of Himachal Pradesh. Tourists experience both Tibetan (Kumar and Gangotia 2016) and local Himachali cultures, is also a home to many yoga and healing centres (Singh 2024). Every year, large number of national tourists visit Dharamshala to learn yoga (Singh et al. 2022). International tourists also visit to learn yoga and Buddhism. The peak season for visiting Dharamshala is usually from March to October, and a productive period for locals who are engaged in hotel and hospitality businesses. Kangra is located around 215 km from the summer capital, Shimla (https://www.google.com/maps).

Numerous stories exist on water in the entire Himachal Pradesh. Historical stories reveal how people tackle water shortages during prolonged droughts. This often includes various superstitious practices. Even stories of human and animal sacrifices are also common. Historical stories exist about how the queen's sacrifices were believed to result in the emergence of new springs. Such cultural beliefs are also common in numerous regions around the globe (Khaneiki 2020). These narratives may stem from the limited scientific understanding in the ancient times, when humans had little knowledge of climate uncertainties (Salite 2019). Such stories were integral part of every society. People also have their own traditional medical treatments for any diseases. These cultural practices were part of the traditions of the Kangra Valley dwellers.

The naturally flowing water in the valley has played a pivotal role in enhancing the socio-economic conditions of Kangra Valley dwellers (Baker 2007; Fischer 2017). However, the valley is currently grappling with water scarcity, reflecting an essence of resource scarcity. The River Beas is the only major river that flows at the southern end of the Kangra District, separating Kangra from the adjacent Hamirpur District (Fig. 1.1). While there are numerous major and minor streams flow in the valley.

Although the people of Kangra at large have relied on their water conservation structures for generations, recent climate change, a surge in anthropogenic activities, and a lack of community support have posed serious challenges in maintaining their indigenous structures.

Kangra has been experiencing rapid urbanization since the beginning of the twenty-first century (Singh 2024). In the Dharamshala vicinity, urbanization has substantially changed land use pattern, reduced agricultural land, challenged agricultural production, and jeopardized businesses. Urbanization is inevitable and posing threats to water and food security. Thus, valley dwellers are shifting from agriculture

---

[2] https://en.wikipedia.org/wiki/dharamshala (Assessed on 01–02–2025).

Fig. 1.1 Map of Himachal Pradesh showing major rivers and administrative districts (adopted from https://www.adb.org/; open access)

to tourism.[3] Infrastructure development is accelerating at a faster rate in the entire Himachal Pradesh.[4]

---

[3] https://www.tribuneindia.com/news/himachal/dharamsala-fest-to-boost-tourism-in-kangra-val ley/ (Assessed on 10–02–2025).

[4] https://www.himachalheadlines.com/opinion/himachals-infrastructure-revolution-connecting-remote-regions/ (Accessed on 10–02–2025).

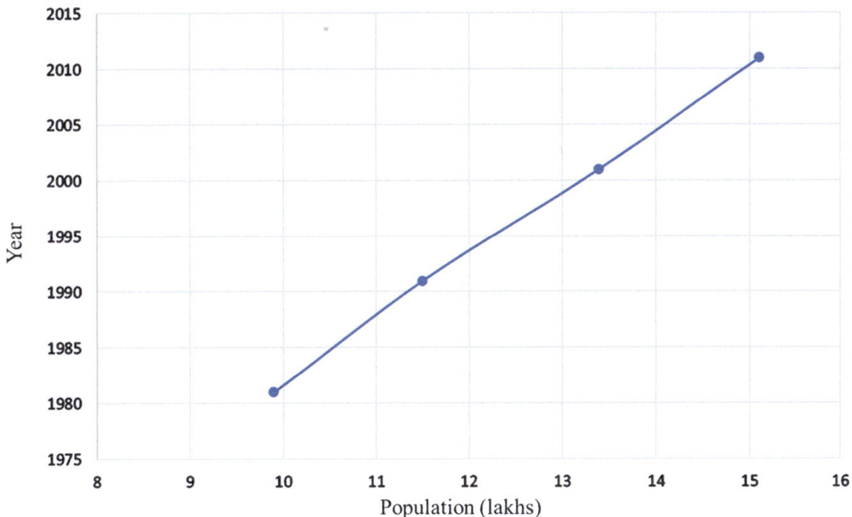

Fig. 1.2 Population growth from 1981 to 2011 in Kangra District of Himachal Pradesh (Indian Census; available at https://censusindia.gov.in/census.website/)

In Kangra Valley, the adoption of modern techniques is replacing all the traditional methods. This shift is quite evident, as modern infrastructure has mostly overtaken traditional construction practices. For instance, locally available materials, such as granite, mud, and wood once commonly used for the construction of houses, and shale rock for roofing, have in large entirely replaced by modern bricks and cement.[5] Even though traditional homes offer several benefits, such as the natural materials keep homes cooler in summers and warmer during winters. However, traditional homes are more expensive and require skilled labour.

In Kangra District, people living in the upper reaches (higher elevation) face many challenges, including extreme cold weather, struggle for water and food, and limited access to transportation, healthcare, and educational facilities. Therefore, population density in the upper reaches is also lower as compared to the lower areas, and at the same time, the life is tough and has many daily life struggles. Despite these challenges, communities in the upper Himalayas have adapted themselves to such harsh climate by optimizing the use of available natural resources for survival. While in the lower reaches, the weather is subtropical to cold temperate, and the large agricultural farms make it a suitable place to survive.

However, with the rapid population growth, the demand for water and food has increased exponentially. The population growth in Kangra is presented in Fig. 1.2.

Most of the Kangra Valley dwellers are engaged in agriculture, which is largely vulnerable to climate change (Datta et al. 2022). Over 80% of farmers are primarily rely on rainfall for irrigation, facing loss to agricultural productivity during weak

---

[5] https://www.tribuneindia.com/news/himachal/kangras-hill-architecture-fading-away-amid-mod ernisation/ (Assessed on 10–02–2025).

monsoons, and prolonged dry spells. Such high dependency on rain-fed agriculture necessitates special attention to face such climate challenges with resilience because climate change often brings extraordinary crises. These stresses further exacerbate the risk of conflict and may potentially fuel instability (www.un.org/en/).

Kangra District of Himachal Pradesh has the highest population in the state and the fourth largest in terms of area. Kangra Valley is one of the most beautiful valleys of Himachal, situated in the foothills of Himalayas and the focus of this book. The valley is known for its rich culture and heritage in the Global South. This is quite reflected from valley dweller's unique traditional music and dances, and their adorable attires, and festivals celebration (Balokhra 1995). Kangra has its own dialect of the Hindi however, Kangri is widely spoken (a blend of Pahadi and Punjabi).

As mentioned, the climate in the valley greatly varies with elevation and season. Summers are usually mild, while winters are colder than the plain areas. Air pollution is relatively low, and the insolation is high due to the clear, and unpolluted atmosphere. Rainfall is often intense and sometimes prolonged. For instance, Dharamshala receive the highest rainfall, and is considered as the second wettest place in the country. Therefore, both of these areas are well-suited for tea plantations.

Kangra Valley is largely susceptible to high-intensity earthquakes. Nearly a century ago, an earthquake in 1905 destroyed numerous infrastructures (The Kangra Earthquake of April 4, 1905). The devastation was so powerful that no major infrastructure was developed by colonial powers afterward (Baker 2001). Since then, the glory of the valley has also diminished. This further restricts any infrastructure and agricultural development. The agricultural sector was particularly affected, and the valley witnessed intermittent development until independence.

After independence, the country had numerous economic challenges, and therefore the advancement in the agricultural sector during the initial years was sluggish. Despite this, Kangra Valley speedily regained momentum when the Himachal Pradesh separated from the adjacent Punjab state and was recognised as a separate state in 1971 (Balokhra 1995). After state recognition, there was a surge to redevelop infrastructure, and other sectors, including agriculture. The valley is intensively cultivated however, current climate change has significantly impacted water resources, raising potential agricultural crises for local communities. Even though the valley contributes negligibly to global warming, it remains vulnerable to notable climate change.

Colonial power showed little interest in the traditional knowledge of the Kangra Valley and their concern was only limited to revenue collection. Hence, colonial power began documenting the Kuhl system of the valley in Riwaj-I-Abpashi (see Chap. 3) to get control and generate revenue.

Nearly 90% of Himachal Pradesh population lives in the rural areas, and more than 60% engaged in agriculture (www.agriculture.hp.gov.in). Thus, agriculture is a major source of livelihood for valley dwellers. Although agriculture in the Kangra Valley is always challenging and faces significant insecurities, primarily due to its high dependency entirely on monsoonal rainfall, which usually brings large uncertainties if longer dry spells exist (Prasad et al. 2023). Farms in the valley are largely nourished with stream water, with negligible reliance on groundwater for irrigation. Only a few farmers have their own tube wells installed with private entities.[6] However, high installation costs and limited irrigable areas restrict its use. Therefore, marginalized communities remain more vulnerable to any uncertainties (Dagdeviren et al. 2021; Gaurav et al. 2025; Sithirith et al. 2024).

Since a large population of the valley is engaged in agriculture, providing full institutional support to all farmers is challenging, even after support from international organizations. Eventually, the farmers have to look into the available infrastructures needed to avail the scheme benefits. For instance, these schemes often have a few defined parameters to get the benefits, which is often quite challenging for the marginal, and tenant farmers. Therefore, such schemes unintentionally favour high-income individuals over others, and often perpetuate inequality, increases debts, and crises. It is quite likely that the privileged families with large landholdings are largely benefited from any scheme, while others remain questionable (Rosencranz et al. 2022). Kafle et al. (2024) found that in developing countries, farmers with stronger social networks are likely more benefitted than poorer ones from any subsidies/schemes. My interaction with farmers in the valley also concluded with the same thoughts. However, no agricultural scheme is designed in such a way that fits for all. Therefore, valley dwellers are seeking alternative off-farm livelihood possibilities. For instance, people are now increasingly looking for employment opportunities in both public and private sectors, while the public sector being the first choice due to lower uncertainties.

Residents of the valley often seek off-farm employment opportunities and move to nearby major cities, such as Chandigarh and Delhi. There is a growing migration trend in search of better opportunities to nearby metro cities, with Chandigarh being the first preferred destination due to the closer proximity over Delhi. This often increased the workload on other family members. In recent years, migration of skilled workers to Middle-Eastern countries, more specifically to Saudi Arabia, and United Arab Emirates (UAE) has also become common. This is quite obvious as it is very challenging for any state to provide employment opportunities to all. Therefore, self-startup for creating employment is strongly encouraged. Some valley dwellers have also migrated to the adjacent Kullu District, which is more advanced, and geographically wider than Kangra Valley. Being on the route to Leh (Ladakh), Kullu has the privilege to attract more tourists than Kangra and is relatively more developed.

Institution's role is crucial in formulating strategies to enhance agricultural productivity, ensuring access to appropriate markets, securing good prices for farmer's products, and simplifying the ease of doing business. Furthermore, institutions are responsible for supporting farmers by providing essential and required training including needed infrastructures. However, many times the infrastructure provided to the farmers are partial due to financial constraints. Thus, the government seeks assistance from international funding agencies to fulfill these needs. Although these agencies provide financial support, they often keep themselves aside

---

[6] Only those farmers who can afford them. Although, tube-wells are not so common in the valley and further subjected to topography, and groundwater availability.

from assisting with the necessary infrastructure for effective, hands-on training to the farmers.

Furthermore, the construction of large dams such as Pong and Bhakra has inundated most of the potential irrigated lands in the regions. This has largely reduced agricultural productivity in the Kangra District, and neighbouring areas. Moreover, the displacement caused by these projects adversely affected the mental health of migrants, with first-generation migrants largely encountered considerable distress.

This study followed a systematic search to collect literature on traditional knowledge along with agro-economy of Kangra Valley. The methodology adopted for this study is described below.

1.2   Methodology for Literature Collection

This study first followed an initial systematic review to gather literature pertinent to the traditional knowledge and agro-economy of the Kangra Valley. Searches were conducted across various databases such as ScienceDirect, Google Scholar, and random searches in Google. However, literature specifically focusses on the topic was limited.

Fieldworks were conducted intermittently between 2014 and 2024 in, and around the Kangra Valley. Data on traditional water management practices were gathered from the farmers who engaged in agriculture for decades. Information was collected from the senior citizens living in the valley for many years, including Patwaris (village accountant), Kohlis (channel water supervisor), Tehsildars (Administrative Officers of a Tehsil), and their families. These individuals possess extensive knowledge of traditional indigenous knowledge of the Kangra Valley.

1.3   Water Conservation Structures and Their Importance in the Kangra Valley

Kangra Valley is home to several traditional water conservation structures, namely, *Kuhl, Boweris, Khatris*, etc., and discussed in the book in details in Chap. 2. In Chap. 3, I briefly discuss on Kuhl system- a unique irrigation method developed centuries ago. This gravity channel irrigation method in the valley supported in growing agricultural productivity for many years.

Water conservation structures such as *boweris* were constructed to collect water in regions where springs discharge is low. These *boweris* served as an important source for drinking. Similarly, *khatris* are small storage water tanks, also built to capture rainwater for drinking and other domestic usage. Although *khatris* are less common in the Kangra Valley, they are more prevalent in the adjacent Hamirpur District. In the lower reaches of the valley where springs are undeveloped, dug-wells are commonly constructed to collect groundwater (see Chap. 2).

Sharma et al. (2015) emphasized the significance of community engagement for the sustainability of Kuhls. However, over time, community interest in managing natural resources has diminished with many indigenous heritages are on the verge of collapse. For instance, Kuhls were developed to irrigate the farms without institutional assistance, are now dying. Losing such heritage is an essence of poverty, conflicts, and insecurity toward water and food.

This book also examines the outcomes of traditional indigenous knowledge. For instance, *gharats* (water mill) were developed in the valley to grind grains. These *gharats* utilize the kinetic energy of Kuhl or stream water flow (Malik et al. 2022; Slariya 2013). Water driven *gharat* offers several advantages over electric mills; (1) the flour produced is more nutritious and has a longer shelf life. (2) save energy by utilizing nature-based solutions; (3) generate no carbon emissions, relying solely on the kinetic energy of water. Therefore, a win–win and sustainable solution for all.

1.4 Spiritual Significance of Water Conservation Structures

Water bodies hold profound religious and cultural significance in the Kangra Valley. Many rituals and festivals are intricately linked to water and weather. For instance, the **Sair** festival, celebrated across Himachal Pradesh, marks the end of Kharif harvest season just after the monsoon. Similarly, the **Lohri** festival signifies the beginning of wheat harvesting and the end of winter. These festivals, along with others observed in the state, are largely associated with seasonal transitions such as winter, summer, and monsoon (Balokhra 1995).

1.5 Agricultural Productivity in the Kangra Valley

Water is essential for enhancing crop yields and soil productivity. However, providing water to farms is challenging in the mountain regions such as Kangra Valley. The valley's limited groundwater resources necessitate the reliance on available surface water primarily derived from glacier melt in the Dhauladhar mountain range. Nevertheless, unpredictable climate change, such as prolonged droughts, erratic rainfall, and dry winter spells, feeds less water to the streams reduces water availability for irrigation. This is further compounded by anthropogenic activities, such as building major, and minor hydropower projects, consequently affecting streams water flow. This adversely affected agricultural productivity in the valley. For instance, there has been a notable decline in paddy production, a crop that requires substantial water over others.[7]

Agriculture in the Kangra Valley is highly susceptible to climate change, primarily because agricultural productivity is largely dependent on water availability. Since the majority of the valley dwellers are engaged in farming for their livelihoods, low productivity from the farms is an essence of exacerbating poverty and insecurity. Ali and Talukder (2008) and Mondaca-Duarte et al. (2020) have demonstrated how reduced irrigation limited crop productivity. This certainly directly impact thousands of farmers and their families. Furthermore, low productivity will result in unemployment and raised prices and exacerbate unnecessary conflicts.

Kangra Valley is largely shaped by khads, the local channels through which water flows. The crystal-clear cold water of khads have carved the valley over thousands of years. The khads water have developed extensive agricultural land, making the region fertile and suitable for cultivating crops. The valley is also rich in biodiversity and home to a wide variety of faunas and floras. Notably, Pong Dam, located in the lower reaches of Kangra is enriched in biodiversity (Sharief et al. 2018). Fishing is a key livelihood for communities living near to the dam. Additionally, during the dry season, when dam water levels recede, newly exposed land is used for sowing winter crops, and harvested before the dam water rises again in summer.

Kangra Valley once received the highest rainfall in Himachal Pradesh, making it more suitable for a variety of crops, including paddy, which requires substantial water and also suitable for tea cultivation. In an estimate, over 2500 litres of water is required to produce 1 kg of rice (Ali et al. 2023), and nearly the same amount of water is needed for 1 kg of tea leaves. These water-intensive crops are primarily irrigated by Kuhl water in the dry season, and also during weak monsoon. The Kuhl water enhances agricultural productivity, supports local growing businesses, and improves socio-economic conditions of the valley's dwellers. However, in recent years, notable climate changes such as prolonged dry spells, and intense floodings have affected every resource including irrigation. Kuhls now receive less water than before, leading to noticeable decline in the paddy cultivation (Sharma et al. 2015). The shift poses a significant threat to both water and food security. Ultimately, with declining local production, the demand for the staple foods will be met by import from other neighbouring states, which could result in higher food prices.

Staple foods such as paddy, wheat, and maize are an essential part of valley dweller's local diets. Their productivity primarily depends on rainfall and often adversely suffers during prolonged dry spells, leading to significant crop failures (Prasad et al. 2023). Additionally, unlike the adjacent Punjab plains, where agriculture is largely relied on groundwater for irrigation, and has achieved high levels of production, agriculture in Himachal Pradesh is mostly rain-fed and thus largely vulnerable to climate change (Lapworth et al. 2014).

Furthermore, orchards farming particularly apple cultivation in the valley is underdeveloped compared to other districts of Himachal Pradesh such as Shimla,

---

[7] *Statistical Year Book Himachal Pradesh 2022-2023.* Retrieved September 22, 2024. Available at https://himachalservices.nic.in/economics/pdf/YearBook2022-23.pdf.

Kinnaur, and Kullu. Other orchards[8] are also well developed in these districts. Kangra Valley receives heavy rainfall during monsoon season (July to September), are more favourable for crops like paddy, lychee, and tea cultivation.

Tea gardens were once largely flourished in the Kangra Valley prior to the devastating 1905 earthquake. Since, the tea industry in Kangra has declined, it is no longer as popular as Northeast India, and Bengal (Pasha et al. 2024). Over time, tea production has plummeted.[9] There could be many reasons for the decline, which could be cited here. Although developed by colonial powers, tea cultivation lost its glory following the 1905 earthquake, and the colonial powers showed little interest in its revival. Another critical factor includes lack of effective marketing and advertising strategies (Comanor and Wilson 1974). Some other challenges are related to product quality, production costs, and limited promotion efforts. All these factors collectively diminished the status of Kangra tea in the Indian markets.

1.6 Other Livelihood Opportunities in the Kangra Valley

The off-farm income opportunities in the Kangra Valley remain limited. Although numerous valley dwellers are also engaged in the growing tourism sector, youths tend to prefer towards government jobs due to their lower uncertainties, and security. Furthermore, tourism in the valley has flourished, especially after the onset of COVID-19 pandemic, spike adoption of work from home culture in the country. However, Kangra remains a less preferable tourist destination compared to nearby localities such as Kullu and Manali.

The valley has also undergone notable land use changes in recent years. Many dwellers are now shifting from farming and increasingly focusing on constructing hostels, hotels, and resorts to capitalize on tourism. Thus, the agriculture is largely dwindling and further compounded by uncertainties associated with climate change.

1.7 Rationale of This Study

The book explores various traditional water heritage systems emphasizing their importance and future challenges. It also examines the role of institutions in promoting traditional practices within the valley. Through this book, I have highlighted several key pressing issues and discussed growing water challenges within the valley. However, there is a clear gap in understanding and integrating traditional

---

[8] Generally, apples in large, which have high market value and demand compared to other orchard fruits in India.

[9] The Tribune "https://www.tribuneindia.com/news/himachal-calling/himachal-calling-brewing-trouble-farmers-look-up-to-govt-for-kangra-tea-revival/" (Accessed on 21–10–2024).

knowledge systems of the Kangra Valley in the modern era, where other new challenges threaten the sustainability of traditional heritages of the region. Thus, this study was undertaken to cover knowledge gaps, focusing on the following crucial points:

- Future of indigenous traditional knowledge in the Kangra Valley.
- Traditional water conservation structures and their role in valley dweller's livelihoods.
- Community preparedness towards climate change uncertainties.
- Is Himachal prepared to pay subsidies to water-resilient crops and/or advanced irrigation schemes?
- Is micro-irrigation a real solution to agricultural enhancement in the Kangra Valley? Why are farmers reluctant in shifting to micro-irrigation?
- Farmers willingness to continue agriculture or shifting to alternate occupations?
- Challenges hindering the implementation of best water management practices in the valley.
- The role of institution's in fostering sustainable development of the Kangra Valley.

The book is not confined to these points, but I also looked into the answers through extensive field works. My interactions and discussions with farmers, seniors, and knowledgeable elders, and the younger generation on various socio-economic issues have been profoundly enriching. These experiences deepen my understanding and encouraged me to explore the traditional indigenous knowledge systems of the Kangra Valley. I gained extreme knowledge while completing this book however, I used my wisdom to arrive at my conclusions.

The book presents a brief overview of current issues and explores how traditional indigenous knowledge, integrated with well-structured agro-economy policies, can transform agriculture and improve the socio-economic conditions of farmers and their families. A particular emphasis is given on the preservation of traditional water conservation systems, highlighting their critical role in ensuring the long-term sustainability of the Kangra Valley.

1.8 Book Outline

The book comprises five chapters, including this introductory chapter. This chapter (Chap. 1) introduces readers to the water crisis issue and agriculture challenges in the Kangra Valley, where agriculture remains the primary source of livelihood. The chapter discusses the significance of water conservation structures and their importance in sustainable livelihoods. Furthermore, the chapter offers an overview of the other chapters, providing clear picture of the book's structure.

Chapter 2 delves into the various traditional water conservation structures, such as *Kuhl, Boweris, Khad, Khatri*, and other structures. The chapter showcases the indigenous wisdom of valley dwellers in developing these systems to harness water for both individuals and agriculture. The chapter also highlights how this water systems fostered societal harmony and their importance in sustainability for generations.

Furthermore, Chap. 3 explored Kuhls in greater detail. The chapter examines Kuhl historical and socio-economic importance, and their role in communities' livelihoods. The chapter highlights community managed assets driven by community cooperation. Chapter 4 focuses on the role of institutions in fostering sustainable development of the Kangra Valley. Chapter 4 outlines the responsibilities of the institutions and underscore the need for refinement of policies so that farmers can avail maximum benefits and support to enhance their socio-economic conditions.

The final chapter (Chap. 5) addresses the pressing challenges and the way forward towards achieving the goal of sustainable development of the Kangra Valley. The chapter proposes strategies for Kangra Valley sustainable development and emphasizes the crucial role of water management in enhancing the socio-economic conditions of the local communities.

Acknowledgements I extend my gratitude to the local people of Kangra Valley, whom I interacted with and their views on various socio-economic aspects were invaluable. Their critical insights on water, agriculture, and other social, and economic issues enriched my understanding and help me to better visualize the valley challenges. I am thankful to those who generously shared their views with patience, which were invaluable to this book.

References

Agarwal M, Sharma M, Shekhar Ganesha S (2024) Gharat—a sustainable source of livelihood. Mater Today: Proc 99:1–7. https://doi.org/10.1016/J.MATPR.2023.04.015

Ali MH, Talukder MSU (2008) Increasing water productivity in crop production—a synthesis. Agric Water Manag 95(11):1201–1213. https://doi.org/10.1016/J.AGWAT.2008.06.008

Ali S, Mehri F, Nasiri R, Limam I, Fakhri Y (2023) Fluoride in raw rice (Oryza sativa): a global systematic review and probabilistic health risk assessment. Biol Trace Element Res 202(9):4324–4333. https://doi.org/10.1007/S12011-023-04004-4

Ali I, Shah AA, Alotaibi BA, Ali A, Khan M (2025) Smallholder perceptions of climate change in high-altitude farming: the influence of household characteristics and local knowledge on adaptation strategies in Nagar District, Pakistan. Clim Services 38:100581. https://doi.org/10.1016/J.CLISER.2025.100581

Aliabadi V, Ataei P, Gholamrezai S (2022) Farmers' strategies for drought adaptation based on the indigenous knowledge system: the case of Iran. Weather, Climate, Society 14(2):561–568. https://doi.org/10.1175/WCAS-D-21-0153.1

Asian Development Bank (2010) Climate change adaptation in Himachal Pradesh. 78. https://www.adb.org/

Baker M (2001) The politics of knowledge: the case of British colonial codification of "customary" irrigation practices in Kangra, India. Himalaya, Journal of the Association for Nepal and Himalayan Studies 21(2). https://digitalcommons.macalester.edu/himalaya/vol21/iss2/7

Baker JM (2007) The Kuhls of Kangra: community-managed Irrigation in the Western Himalaya. University of Washington Press (Verlag). ISBN: 978-0-295-98764-4.

Balokhra JM (1995) The wonderland Himachal Pradesh: a survey of the geography, people, history, administrative history, art and architecture, culture, and economy of the state. H.G. Publications.

Comanor WS, Wilson TA (1974) Advertisement and market power. 256. https://www.hup.harvard.edu/books/9780674005808

Dagdeviren H, Elangovan A, Parimalavalli R (2021) Climate change, monsoon failures and inequality of impacts in South India. J Environ Manage 299:113555. https://doi.org/10.1016/J.JENVMAN.2021.113555

Darwish T, Shaban A, Faour G, Jomaa I, Moubarak P, Khadra R (2024) Transforming irrigated agriculture in semi-arid and dry Subhumid Mediterranean conditions: a case of protected cucumber cultivation. Sustainability 16(22):10050. https://doi.org/10.3390/SU162210050

Datta P, Behera B, Rahut DB (2022) Climate change and Indian agriculture: a systematic review of farmers' perception, adaptation, and transformation. Environ Challenges 8:100543. https://doi.org/10.1016/J.ENVC.2022.100543

Fischer HW (2017) Harnessing the state: social transformation, infrastructural development, and the changing governance of water systems in the Kangra district of the Indian Himalayas. Ann Am Assoc Geogr 107(2):480–489. https://doi.org/10.1080/24694452.2016.1232614

Gaurav K, Sharma P, Sharma A (2025) Revisiting community-based traditional irrigation systems in India. Heliyon 11(1):e41684. https://doi.org/10.1016/J.HELIYON.2025.E41684

Hansen CJ, Currell MJ, Flynn E (2024) Iman peoples water sovereignty: extractive industries in Central Queensland. Extractive Industries Soc 20:101560. https://doi.org/10.1016/J.EXIS.2024.101560

Kafle K, Balasubramanya S, Stifel D, Khadka M (2024) Solar-powered irrigation in Nepal: implications for fossil fuel use and groundwater extraction. Environ Res Lett 19(8):084012. https://doi.org/10.1088/1748-9326/AD5F46

Khaneiki ML (2020) Cultural dynamics of water in Iranian civilization. In: Cultural dynamics of water in Iranian civilization. Springer International Publishing. https://doi.org/10.1007/978-3-030-58900-4

Krampe F (2016) Water for peace? Post-conflict water resource management in Kosovo. 52(2):147–165. https://doi.org/10.1177/0010836716652428

Krampe F, Hegazi F, VanDeveer SD (2021) Sustaining peace through better resource governance: three potential mechanisms for environmental peacebuilding. World Dev 144:105508. https://doi.org/10.1016/J.WORLDDEV.2021.105508

Kumar S (2023) Sustainable rural tourism in Himalayan foothills: environmental, social and economic challenges: a study of Himachal Pradesh. In: Sustainable rural tourism in Himalayan foothills: environmental, social and economic challenges: a study of Himachal Pradesh. Springer International Publishing. https://doi.org/10.1007/978-3-031-40098-8

Kumar A, Gangotia A (2016) Understanding the level of participation of Tibetan community in tourism developmental activities in Kangra district Himachal Pradesh. Tourism Spectrum 2(2):69–76

Lapworth DJ, Gopal K, Rao MS, MacDonald AM (2014) Intensive groundwater exploitation in the Punjab: an evaluation of resource and quality trends. British Geological Survey, Nottingham, UK, 34 pp. (OR/14/068, unpublished)

Malapane OL, Chanza N, Musakwa W (2024) Transmission of indigenous knowledge systems under changing landscapes within the Vhavenda community, South Africa. Environ Sci Policy 161:103861. https://doi.org/10.1016/J.ENVSCI.2024.103861

Malik G, Saini A, Singh H (2022) Gharats: ancient technological marvel at the verge of extinction. ECS Trans 107(1):6933–6943. https://doi.org/10.1149/10701.6933ECST

Mondaca-Duarte FD, van Mourik S, Balendonck J, Voogt W, Heinen M, van Henten EJ (2020) Irrigation, crop stress and drainage reduction under uncertainty: a scenario study. Agric Water Manag 230:105990. https://doi.org/10.1016/J.AGWAT.2019.105990

Negi B, Negi VS, Rana SK, Bhatt ID, Manasi S, Nautiyal S (2025) Role of traditional ecological knowledge in shaping climate resilient villages in the Himalaya. J Environ Manage 376:124325. https://doi.org/10.1016/J.JENVMAN.2025.124325

Pasha SV, Dadhwal VK, Kumari K, Ali N (2024) Historical expansion of tea plantations over 150 years (1876–2023) in North Bengal, India. Environ Monit Assess 196(11):1–18. https://doi.org/10.1007/S10661-024-13208-7

Prasad P, Gupta P, Belsare H, Mahendra CM, Bhopale M, Deshmukh S, Sohoni M (2023) Mapping farmer vulnerability to target interventions for climate-resilient agriculture: science in practice. Water Policy 25(8):815–834. https://doi.org/10.2166/wp.2023.036

Rautela P (2015) Traditional practices of the people of Uttarakhand Himalaya in India and relevance of these in disaster risk reduction in present times. Int J Disaster Risk Reduction 13:281–290. https://doi.org/10.1016/J.IJDRR.2015.07.004

Rosencranz A, Puthucherril TG, Tripathi S, Gupta S (2022) Groundwater management in India's Punjab and Haryana: a case of too little and too late? J Energy Nat Resour Law 40(2):225–250. https://doi.org/10.1080/02646811.2021.1956181

Salite D (2019) Explaining the uncertainty: understanding small-scale farmers' cultural beliefs and reasoning of drought causes in Gaza Province, Southern Mozambique. Agric Human Values 36(3):427–441. https://doi.org/10.1007/S10460-019-09928-Z

Sharief A, Paliwal S, Sidhu AK, Kubendran T (2018) Studies on bird diversity of pong dam wildlife sanctuary, Kangra, Himachal Pradesh, India. J Entomol Zool Stud 6(4):904–912

Sharma AK, Sharma KD, Prakash B (n.d.) Death of Kuhl irrigation system of Kangra valley of Himachal Pradesh: institutional arrangements and technological options for revival. Ind J Agri Econ 70(3)

Sharma AK, Sharma KD, Prakash B (2015) Death of Kuhl irrigation system of Kangra valley of Himachal Pradesh: institutional arrangements and technological options for revival. Indian J Agric Econ 70(3):350–364. https://doi.org/10.22004/AG.ECON.230213

Simangan D, Lee CY, Sharifi A, Lee Candelaria J, Kaneko S (2022) A global analysis of interactions between peace and environmental sustainability. Earth Syst Govern 14:100152. https://doi.org/10.1016/J.ESG.2022.100152

Singh U (2024) In-between metropolitan cities and urban theories: a case of small town Dharamshala. GeoJournal 89(1):1–16. https://doi.org/10.1007/S10708-024-10996-W

Singh U, Upadhyay SP, Jha I (2022) The co-production of space in a tourist city: a case of Dharamshala. Cities 131:103998. https://doi.org/10.1016/J.CITIES.2022.103998

Sithirith M, Sao S, de Silva S, Kong H, Kongkroy C, Thavrin T, Sarun H (2024) Water governance in the Cambodian Mekong delta: the nexus of farmer water user communities (FWUCs), community fisheries (CFis), and community fish refuges (CFRs) in the context of climate change. Water 16(2):242. https://doi.org/10.3390/W16020242

Slariya MK (2013) Hydroelectric power projects-A threat to existing traditional knowledge: a study of power projects in Ravi Basin in Chamba district of Himachal Pradesh India. Asian J Multidimens Res (AJMR) 2(3):65–78.

Unfried K, Kis-Katos K, Poser T (2022) Water scarcity and social conflict. J Environ Econ Manag 113:102633. https://doi.org/10.1016/J.JEEM.2022.102633

Zaryab A, Nazari A, Farahmand A, Yaqubi MS, Mirzad SMM, Jafari Z, Ibrahimi MT, Zaki H, Ali S, Shams AK (2025) Issues and challenges in sustainable usage of groundwater resources in Afghanistan, pp 259–277. https://doi.org/10.1007/978-3-031-79122-2_11

Chapter 2
Indigenous Water Conservation Structures in the Kangra Valley of Himachal Pradesh

Abstract The chapter explores various traditional water conservation structures of the Kangra Valley in Himachal Pradesh. These structures were built by the local communities using indigenous wisdom passed through generations and adaptability to face any harsh conditions. They also set a unique example of community participation in managing their natural resources sustainably. However, over time, these structures are losing their importance, facing lack of communities' participation and thus are under unprecedented pressures. These indigenous structures are important as they not only provide water but are also imperative to combat water and food scarcity, necessitating their significance amid current challenges posed by unpredictable climate change. This chapter examines the existing traditional water conservation practices in the Kangra Valley and their role in valley dweller's livelihoods.

Keywords Water Conservation Structures · Indigenous culture · Kuhl · Springs · Kangra Valley

2.1 Introduction

The Himalayas, often considered as a "Third Pole" due to their extensive glacier cover, provide substantial freshwater and feed numerous natural streams, and springs (Zhang et al. 2019). In Global South, water is an important part of the culture and plays a crucial role in fostering harmony, peace, and shaping the cultural identities (Johnston et al. 2012; Khaneiki 2020). Water and food are an integral part of the Kangra Valley culture.

Located in the midst of the Himalayas, the Kangra Valley has many remarkable natural wonders such as high Dhauladhar snow covered mountain ranges, crystal clear streams water (*khad*), various types of rocks, cool and calm weather, clean atmosphere, and picturesque landscapes. These natural wonders make the valley a popular destination among national and international tourists.

The Kangra Valley is endowed with numerous natural water resources that have nourished local communities for generations. Traditional water systems in the valley

S. Ali, *Traditional Water Conservation Community-Managed Structures and Their Role in Valley Dwellers' Livelihoods*, SpringerBriefs in Water Science and Technology, https://doi.org/10.1007/978-3-032-04637-6_2

are not merely water structures, they are deeply embedded in the culture and traditions that brings an essence of peace and harmony among valley dwellers. Currently, the valley is grappling with water scarcity, reflecting a broader essence of resource scarcity. Although the people of Kangra have survived on traditional water conservation structures because they provided meaningful results. Recent climate change[1] and declining community participation highlight the pressing need to promote integrated management of water resources. This is essential for achieving self-reliance and addressing the escalating crises of water and food insecurity.

Kangra Valley is highly vulnerable to frequent and severe climate extremes, reflecting the urgent need to ensure water and food resources. To address these urgent concerns, traditional water conservation structures play a pivotal role in the prosperity of the valley dwellers. Hence, the valley is home to several indigenous water conservation structures such as *Kuhls, Boweris, Gharat* (watermill; outcome of indigenous knowledge), *Khatris*, and *Dug Wells*. These indigenous structures exemplify community-driven action in accessing and managing their water resources. This chapter examines the natural water system that exists in the valley and is briefly discussed in the subsequent sections. Additionally other natural water resources such as khads, springs, waterfalls, lakes, and hot springs are also presented. These resources are important for their ecological and cultural significance.

2.2 Traditional Water Conservation Systems in the Kangra Valley

In the Kangra Valley, local communities have built several traditional water conservation structures for their survival and livelihoods. This section discusses traditional structures like *Kuhls, boweris, gharat, khatris*, and *dug-wells*. Currently these valuable resources are losing their significance, facing non-cooperation from the communities, and are on the verge of dying.

2.2.1 Kuhls

Kuhls are gravity-driven water channel systems constructed by the Kangra Valley dwellers to supply adequate water for irrigation during the dry season and/or weak monsoon (see Chap. 3 for detail). The unique water flow systems can nourish over thousands of hectares of land within their command areas. Kuhls reflect community wisdom and used to nourish the farms for over a century. This enhances agricultural productivity by enabling farmers to cultivate multiple crops annually. The Kuhl system is managed entirely by local communities without any central intervention.

---

[1] https://www.tribuneindia.com/news/himachal/dry-spell-hits-drinking-water-irrigation-schemes-in-kangra/ (Assessed on 10–02–2025).

Although not all farms are connected with the Kuhls, however most of the farms are benefitted from them. Currently, only limited Kuhls receive a substantial water supply (see Chap. 3).

2.2.2 Boweris

Boweris are traditional step-type water conservation structures, generally rectangular in shape, built to collect low discharge spring water. The water is primarily used for drinking and other domestic purposes and served as a vital water source for rural communities. Most villages in the valley have one or two boweris (even more) in every village vicinity. Boweris were once an important source of drinking for many years, until the pipe water was introduced into the valley. Boweris are often covered by rooftop, reflecting the collective efforts from local communities in the region. The boweris have been an important source for drinking and empowered valley dwellers over generations (Sharma and Kanwar 2009; Fig. 2.1).

Boweris continued to be considered as a reliable drinking source, particularly in the rural areas of the Kangra Valley and have played a pivotal role in empowering local communities. However, recent climate change and rapid urbanization have endangered traditional water conserving structures including boweris. This necessitates timely intervention and combined rejuvenation efforts from communities and institutions to save cultural heritage and traditional knowledge of the valley. Furthermore, boweris remains a vital resource to fulfill the growing demand for water in the region.

Fig. 2.1 Figure shows a boweri in the upper Dharamshala. *Note* they are protected by rooftop and also a sacred place for the local people. Also note that a small pipe is connected to drain the spring water, which was earlier missing, reflecting present engineering as collected water requires periodic cleaning of the boweris

Boweris are regularly maintained by communities without any financial support. They are regarded as a sacred place and people avoid wearing footwear in its vicinity. Usually, small temples can be found near boweris (Fig. 2.1), underscoring their deep respect for their water resources, which provided water to them for drinking and other domestic purposes.

Presently, boweris are fading out due to the widespread availability of tap water connections provided to every household under various government schemes. However, the tap water often unreliable among valley dwellers due to frequent supply disruptions. The primary challenges lie in the water supply system is regular maintenance of pipelines, often hindered by the lack of infrastructure, inadequate funds, and other local constraints. The situation becomes even worse when tap water is unavailable for prolonged periods.

Therefore, natural water resources such as springs remains a reliable and cost-effective solution, necessitating their rejuvenation and protection in the Kangra Valley. In India, maintaining the proper infrastructure under any centralized schemes is always challenging. Thus, traditional and natural water resources should not be overlooked in favour of modern advancements, instead both should be integrated to strengthen the community resilience and enhancement of valley dweller's livelihoods.

Boweris are the prime example of traditional wisdom passed down through generations. It is estimated that over hundreds of boweris exist across the Kangra Valley with varying discharges. Their maintenance is entirely managed by local communities, without any central interventions. Eventually, uncontrolled rapid urbanization, land degradation, and climate change consequently affected the region's springs, rendering numerous boweris dry and unmaintained.

2.2.3 Gharat

Gharats are traditional watermills were once an integral part of the culture of communities in the Western Himalayas (Bhatt et al. 2021; Kumar et al. 2018; Malik et al. 2022; Fig. 2.2). These watermills operate using kinetic energy of flowing water, sourced from Kuhl or streams. Gharat has several benefits and discussed in Chap. 1. For instance, the generated flour from gharat is more nutritious than that from electrically driven mills, due to the low speed and temperature involved in the grinding process. Currently, many gharats have become defunct, while a few are seasonally operational subject to the availability of water in the Kuhls or streams. Their decline reflects water scarcity and changing socio-economic practices within local communities.

Fig. 2.2 The yellow arrow in the **a** indicates the inlet of Kuhl water used to rotate the gharat. **b**, and **c** shows the front and side views of the gharat, respectively (*Courtesy* Shubi Shubham)

2.2.4 Khatri

Khatri's are man-made structures built to store rainwater and mainly used for drinking purposes. They are usually small water storage tanks constructed in areas devoid of springs. These structures fill with water during the rainy seasons, as rainwater slowly percolates through the existing lithology, and is naturally filtered. Khatris are generally intended for public use, however private khatris were also constructed by the privileged families for their personal usage. These structures unintentionally act as rainwater harvesting systems.

Khatris are less common in the Kangra Valley due to the presence of springs however, they are more in number in the adjacent Hamirpur District. These structures are often enclosed with doors to protect the stored water from wild animals, ensuring the water remains clear for drinking and other domestic uses.

Khatris are typically carved into Shivalik rocks and constructed in such a way that the stored water will not further infiltrate. They were common in ancient times, however many of them are now extinct. Khatris shows how valley dwellers used nature-based solutions to filter rainwater for drinking.

Therefore, both boweris and khatris are important indigenous water storage structures constructed that have sustained communities for generations, demonstrating how traditional wisdom saved local communities to survive over generations. These structures are also imperative for addressing water scarcity, necessitating their importance under current climate change conditions. Currently, these structures are less common and almost defunct.

2.2.5 Dug Wells

Dug wells are commonly found in the lower areas of the Kangra Valley, particularly where springs are absent. These wells play a crucial role in providing water for drinking and other domestic uses (Fig. 2.3). Dug wells were constructed by the local communities to get water where other water sources were unavailable. The wells were constructed manually with no machinery involved. Therefore, usually one well was constructed in a village, serving the entire community's water needs.

With the advent of electric motors, water from dug wells has occasionally been used for limited irrigation, especially in areas where Kuhls were not developed. In ancient times, it was culturally uncommon to purchase food from outside. Perhaps this might be due to low household incomes, as most families lived in rural areas and were economically underprivileged. Thus, communities relied on their local grown food. However, this perception has completely changed in recent years with improved income levels.

Fig. 2.3 Dug well located at the lower reaches of Kangra Valley (*Courtesy* Shubi Shubham)

2.3 Other Local Natural Water Systems in Kangra Valley

2.3.1 Khad

Another important water source in the Kangra Valley is streams locally known as "Khad".[2] Khads are vital livelihood assets for the valley dwellers, and played a pivotal role in shaping the landscape over the years. Khads are perennial streams carrying glacier meltwater to the rivers or reservoirs. However, many khads are also ephemeral, meaning they receive water only during monsoon season and dry up in other seasons. The water carried by streams provides substantial water to nearby water bodies, including the Pong Dam and lastly joins the Beas River.

There are numerous major khads in the Kangra Valley including Binwa, Neugal, Baner, Gaj, and Dehar Khad (https://himachal.nic.in/en-IN/index.html). Besides their hydrogeological significance, khads enhances natural beauty of the Kangra Valley. Tourists seeking a spot away from all distractions can visit the beautiful crystal-clear water of khads, where the melodious sound of water adds to the ambience. Therefore, many resorts and restaurants are located along the khads just to have a good ambiance.

However, water level in numerous khads drastically changes, particularly during the monsoon season. Monsoon supercharged by climate change, often brings more water to the khads, leading to destruction of numerous infrastructures, damage agricultural land, and endanger human lives. Also, climate change disrupts regular water supply in numerous khads during dry season, threatens the biodiversity that depends on these water resources and often disrupt water availability to the local communities.

Khads are an important natural water resource that supports local communities and ecosystems in the Kangra Valley. These natural streams support aquatic life, and provide water to nearby Kuhls (see Chap. 3), which are essential for irrigation. The water from khads plays a significant role in enhancing drought resilience for agricultural communities by ensuring water availability during the dry seasons.

In addition, most households receive water for their drinking and domestic purposes are sourced from khads through various government schemes. This reduces dependency on the groundwater resources, which is already limited throughout the valley. The khad water also enhances groundwater recharge. Therefore, check dams have been constructed on a few khads to improve groundwater table (Fig. 2.4).

Currently, khads are exposed to littering, pollution and the disposal of untreated sewage, leading to significant degradation of khad water. This pollution poses significant threat to biodiversity, disrupt human assets, and accentuate health risks (Fig. 2.5).

Khads in the valley create a pleasant ambiance to attract tourists. Therefore, the areas serve as popular picnic spots to the locals and Tibetan monks who spend their weekends during summers and swim in the chilly water for refreshment. Some of

---

[2] Geologically, khads are natural depressions where stream water flows. Loosely the term khad is also referred for streams and is widely used while speaking.

Fig. 2.4 A check dam constructed on a khad in the Lower Dharamshala to recharge groundwater in the downstream (built by the Central Ground Water Board, India (CGWB); the photo was captured by the author in June 2017)

the recreational activities have also developed on khad. For example, a swimming pool has been constructed in the Bhagsu locality of Upper Dharamshala (Fig. 2.6).

2.3.2 Springs

Springs are important in building a water-resilient Kangra Valley and served as a vital resource in shaping the lives of local communities. Local people built boweris (see Fig. 2.1) to collect water from low discharge springs, while high discharge springs (locally known as *Nuadu/Nadu*; Fig. 2.7) are left to flow freely.

Springs are an important water resource in mountains and feed numerous streams. Springs were an important source for drinking until tap water supply was introduced to individual household. Currently this precious water resource is under threat and facing a lack of regular maintenance. This is further exacerbated by the rapid infrastructure development in the region.

Thus, there is an urgent need to revive springs for fostering water resilience in the valley. Springs are the primary source of drinking and provide water to the khads. However, the emerging global climate crisis has adversely affected every natural

Fig. 2.5 Figures show minor khads running out of water even just after monsoon season (July to September) (Photos were captured by the author in November 2024). The lack of water in the khads in recent years have largely affected biodiversity. The khads are also exposed to litter and untreated sewage disposal

resource, including water. Therefore, springs should be developed to maintain their flows however, the book abstains from the technological methodologies of spring development, which could be found in the study of Shrestha (2018) and Pant et al. (2024).

2.3.3 Lakes

Dal Lake is the major lake only situated in Upper Dharamshala, while there are numerous other lakes in the higher elevations of the Dhauladhar ranges (Rai et al. 2024). Dal Lake situated at Dharamshala has experienced receiving less water and large siltation poses a growing threat to aquatic life.[3] Lakes in the mountains attract more tourists and have ecological significance too. They are essential for preserving biodiversity.

---

[3] The Tribune News published on 2 October 2024 "Dal Lake in Dharamsala drying up again, locals launch campaign to save dying fish".

Fig. 2.6 Swimming pool is constructed on the khad water in the Upper Dharamshala (**b**). Promoting water sports in the valley can be clearly visible. The water of the khad is also used for drinking purposes by the locals. *Note* a small temple is constructed along the khad, showing valley dweller's deep respect towards their water resources (**a**)

2.3.4 Hot Springs

A natural hot spring also found at Tatwani village, Lower Kangra. The hot natural spring water has many skin benefits and is traditionally been used for many years. As a result, it often attracts tourists to take baths. Although, it is the only hot spring present in the Kangra Valley. The hot spring also supplies water to the nearby khads.

Fig. 2.7 Nuadu (high discharge spring) in the Lower Kangra Valley (*Courtesy* Shubi Shubham). *Note* a pipe is connected to the springs to ease water collection. Such structures reflect indigenous knowledge of the valley dwellers

2.3.5 Waterfalls

There are three major waterfalls in the Kangra Valley namely Bhagsu, Bangor, and Khabru Waterfall. Bhagsu Waterfall is located in Upper Dharamshala, while Bangoru, and Khabru Waterfall are situated in the Bir Biling and Boh areas, respectively. These waterfalls are an important source of water and supply a substantial amount of water to the khads. Although there are other waterfalls that are seasonal. All waterfalls are approachable by trek and tourists largely visit these areas for the adventure.

2.4   Conclusion

This study reveals that all traditional water structures in the Kangra Valley are under threat of climate change. As a result, farmers are adjusting themselves under the present climate change conditions (Khan and Hussain 2024; Sassi et al. 2024). Similarly, Kangra Valley dwellers are also aware of such changes and farmers are changing their agricultural practices.

In response, valley dwellers are now shifting from agriculture to tourism related businesses. However, in all likelihood, underprivileged communities/marginal farmers and their families are more vulnerable to climate change.

This chapter has explored various traditional and natural water systems that exist in the Kangra Valley. The study suggests that all traditional water conservation structures should be rejuvenated. These traditional heritage water conservation systems exist in the valley, reflecting how communities used their wisdom to cope with present water scarcity and used them for enhancing their socio-economic conditions for generations. These indigenous structures provided meaningful results to the valley dwellers and also a best learning of what others can learn.

These water conservation structures are important in preserving biodiversity, the environment, history, and culture. Therefore, the structures should be conserved along with other feasible development efforts in the valley, necessitating their importance under the present climate change conditions.

References

Bhatt A, Rana D, Lal B (2021) Gharat: an environment friendly livelihood source for the natives of western Himalaya, India. Environ Dev Sustain 23(12):18471–18487. https://doi.org/10.1007/S10668-021-01455-4

Johnston BR, Barber M, Strang V, Klaver I, Hiwasaki L, Castillo AR (2012) Water, cultural diversity, and global environmental change: Emerging trends, sustainable futures? In: Water, cultural diversity, and global environmental change: emerging trends, sustainable futures? pp 1–560. https://doi.org/10.1007/978-94-007-1774-9

Khan MA, Hussain W (2024) Climate change impacts on Pakistan's mountain agriculture: a study on Burusho farmers' adaptation strategies towards livelihood sustainability. In: Traditional knowledge and climate change, pp 21–45. https://doi.org/10.1007/978-981-99-8830-3_2

Khaneiki ML (2020) Cultural dynamics of water in Iranian civilization. In: Cultural dynamics of water in Iranian civilization. Springer International Publishing. https://doi.org/10.1007/978-3-030-58900-4

Kumar R, Paul KVK, Doda N, Jammu S, Charan Sharma B, Peshin R (2018) Traditional water mills (Gharats)-a source of rural livelihood in mountainous region of Jammu and Kashmir. Indian J Traditional Knowl 17(3):569–575.

Malik G, Saini A, Singh H (2022) Gharats: ancient technological marvel at the verge of extinction. ECS Trans 107(1):6933–6943. https://doi.org/10.1149/10701.6933ECST

Pant N, Hagare D, Maheshwari B, Rai SP, Sharma M, Dollin J, Bhamoriya V, Puthiyottil N, Prasad J (2024) Rejuvenation of the springs in the Hindu Kush Himalayas through transdisciplinary approaches—a review. Water16(24):3675. https://doi.org/10.3390/W16243675

Rai SK, Dhar S, Sahu R, Kumar A (2024) Two decades of glacier and glacial lake change in the Dhauladhar mountain range, Himachal Himalayas, India (2000–2020). J Indian Soc Remote Sens 52(3):633–644. https://doi.org/10.1007/S12524-024-01849-7

Sassi M, Ali Y, Ali I (2024) Climate change perception and adaptation among farmers in the mountains of Western Karakoram, Pakistan. Dev Pract:1–18. https://doi.org/10.1080/09614524.2024.2417249

Sharma N, Kanwar P (2009) Indigenous water conservation systems-a rich tradition of rural Himachal Pradesh. Indian J Traditional Knowl 8(4):510–513

Shrestha RB (2018). Protocol for reviving springs in the Hindu Kush Himalaya: a practitioner's manual. International Centre for Integrated Mountain Development. https://lib.icimod.org/

Zhang F, Thapa S, Immerzeel W, Zhang H, Lutz A (2019) Water availability on the third pole: a review. Water Secur 7:100033. https://doi.org/10.1016/J.WASEC.2019.100033

Part II
Water Channels for Irrigation and Their Role in the Livelihoods of Kangra Valley's Dwellers

Chapter 3
Kangra Valley's Kuhl System: The Last Hope for Irrigation in Peril

Abstract This chapter explores the traditional gravity-fed irrigation channels locally known as *Kuhl/Kuhal/Kul*. These dendritic gravity-driven systems were constructed to nourish farms during dry season and also in the wet season when prolonged dry spells prevail in the Kangra Valley. The Kuhl system has played a pivotal role in enhancing crop productivity, supporting local businesses, and sustaining rural livelihoods. The community-built and managed system has provided water for nourishing over thousands of hectares of land in the command areas, set a best model of community-managed natural resources, despite many challenges. The chapter underscores the significance of Kuhls in boosting agricultural productivity and discusses challenges in managing the traditional system, which once supported valley dwellers for many years.

Keywords Community managed system · Kangra Valley · Himachal · Kuhl · Western Himalayas

3.1 Introduction

Water availability for irrigation determines the socioeconomic well-being of farmers and their families. However, water security for irrigation is under unprecedented pressure and facing many challenges under the current extreme climate change condition (Hanjra and Qureshi 2010; Misra 2014).

Although local indigenous knowledge is often underrated over Western technologies (Malapane et al. 2024) however, the local knowledge is the result of many long years of practices into feasible reality (Gómez-Baggethun 2022). This realizes that developing a nature-based solution through generation's wisdom for the enhancement of irrigation is really a successful invention. One such notable example of such innovation is "*Kuhl*" system in the Kangra Valley of Himachal Pradesh[1] (Baker 1996; Coward 1990).

---

[1] https://www.fao.org/4/x5672e/x5672e03.htm (accessed on 23–10–2024).

© The Author(s), under exclusive license to Springer Nature Switzerland AG 2025   35
S. Ali, *Traditional Water Conservation Community-Managed Structures and Their Role in Valley Dwellers' Livelihoods*, SpringerBriefs in Water Science and Technology,
https://doi.org/10.1007/978-3-032-04637-6_3

Kangra Valley is located at an average elevation of 700 m above mean sea level, lack major rivers from which water can be uplifted for irrigation. Even though lifting water from distant sources consumes energy, generates carbon emissions, and is often costly. Thus, the valley can only be irrigated from water flowing in streams. Therefore, local people developed a unique irrigation system called "Kuhls" in which streams water is diverted and used for nourishing farms.

Kuhls are the best traditional and unique way the valley adopted to nourish their agricultural field and supports farming during dry seasons or if precipitation dwindles during monsoon.[2] In other words, Kuhl reduces crop failure during prolonged droughts. These gravity channels also exhibit how community-managed irrigation systems have changed the farming in the mountainous terrain and marked increase in the agricultural productivity (Ashraf and Ahmad 2021; Ashraf and Batool 2019).

Kuhls are the engineering marvels built by the local communities (Fig. 3.1). Kuhls of Kangra Valley is an irrigation system developed hundreds of years ago[3] (Baker 2001a). An essence of food scarcity in prolonged drought years realise to invent other irrigation options instead of relying only on monsoonal rainfall, which occurs during a limited period. Therefore, Kuhls were developed to irrigate farms and to combat food scarcity to enhance annual crop productivity. It is interesting to note here that the whole system is managed by the local communities with no institutional intervention. Such a well-managed undefined irrigation system successfully ran for many years in the valley.[4,5] The traditional water system in the midst of the Himalayas is the best example of adopting nature-based solutions for farming enhancement and self-reliance towards food and water. This also reflects how traditional wisdom has passed through many generations in the valley and gave liberty to valley dwellers to grow multiple crops annually. Since most of the valley dweller's occupation is agriculture, Kuhl has become an important system for enhancing businesses and supporting families.

Kuhls are farmer's owned irrigation systems and were once entirely managed and operated by local communities. Kuhls were designed by the local people's experience to get water for irrigation to their farms. This shows how communities have developed a unique skill to manage their water resources efficiently without any institutional support. One such arrangement can also be traced to Iran where Qanat water is sustainably used by the communities to provide water to the farms (Khaneiki 2020). Similarly, the plains in North India also developed such irrigation arrangements (Gaurav et al. 2025). Pinto (2014) also highlights how indigenous knowledge supports communities on a small East Timor Island.

One of the largest Kuhl systems in the Kangra Valley exists in the Palampur area, draw water from the Neugal Khad (Baker 2001a). The Palampur region is also known for its tea gardens, and thus, many tea gardens that were earlier started by colonial powers were largely inundated by Kuhls water.

---

[2] Monsoon season in India usually starts from July and ends in September.

[3] https://en.wikipedia.org/wiki/Kuhl_irrigation_(Himachal_Pradesh); Accessed on 20–10–2024.

[4] https://www.youtube.com/watch?v=uFCNuzn7Q_8 (accessed on 20–10–2024).

[5] https://www.youtube.com/watch?v=KlbxUAaI5RA (accessed on 20–10–2024).

Fig. 3.1 A Kuhl is carved from a Khad at Khaniyara, Upper Dharamshala, Kangra

The existence of such a well-developed irrigation network system in the valley reflects traditional wisdom of the local people, which passes through generations. The gravity-fed channels have a long and rich history and primarily built by local communities with few instances of support from kingdoms who ruled the area. In this context, more detail can be found in Baker (1996).

3.2 Drivers of Conflicts in the Kuhl System

Although Kuhls have been a lifeline for agriculture in the Kangra Valley, their maintenance and operation were never easy. There were many challenges in effective functioning of Kuhl systems. Some of the major conflicts are listed below:

- Disputes over sharing of water for irrigation
- Regular maintenance of Kuhls
- Gathering of farmers for solving conflicts
- Disrespect towards the Kohlis
- Delays in providing cereals[6] to the Kohlis in return for their services
- Negotiation with privileged and high-caste families who often argued for more water access.

---

[6] Earlier, people gave cereals grown on their farms in exchange for labour.

3.3 Role of Kohli

To resolve any water-sharing conflicts and fair distribution of Kuhls water, local communities recruit a Kohli. Kohli can be considered as a "supervisor of Kuhls". Kohlis are generally selected by local communities who have the skills and willingness to take on this responsible duty. They are usually appointed from the same Kohli family. The Kohlis role was pivotal, who along with local communities, takes care of Kuhls. The Kohlis are skilled negotiators and capable of resolving conflicts over water use with farmers and other Kohlis and ensure each field gets sufficient water while maintaining harmony. They gained these skills from their past generations.

As mentioned earlier, Kuhls water sharing has gone through many conflicts over its allocations. The Kohlis have the wisdom to resolve conflicts by maintaining harmony through negotiation. It is an everyday ongoing struggle that Kohlis (Kuhl supervisor) and communities have to keep proper vigilance and crucial during the monsoon season. The most vulnerable areas during monsoon are steep slopes and sharp bends where the flow of water is high (Fischer 2017).

However, with the decline of the Kuhl system, the role of Kohlis, and community participation are also diminishing. The loss of such tradition is an essence of insecurity, that has long supported local communities.

3.4 State Intervention

After state intervention, several Kuhls, which were earlier constructed using local materials are concretized with an intention to save water loss from infiltration. With the state intervention, few Kuhls are now in control of state government, and others remains under community management. This greatly affected the water distribution and has been a source of contentious for many years. As a result, there is a noticeable decline in the water-intensive crops such as paddy.

Few Kuhls are recently been cemented/concretised by the state institutions. However, there are worries as cemented Kuhl's once broken require substantial funds to get repaired, which often involves complex procedures. This often results in the release of water to the adjacent farms and eventually the farm owners have to bear the cost of repairs (Sethi 2011). Additionally, the concrete structures make it difficult for livestock to access water for drinking.

Another growing concern is the reduced water availability in the Kuhls due to the construction of micro-hydropower plants in upper reaches of the valley, which significantly diverted water. The situation becomes severe during dry summer months, when most of the Kuhls run out of the water (see Fig. 3.2).

Fig. 3.2 Kuhls in the Kangra Valley. Most Kuhls have now been concretized with state intervention. Note the reduced water flow in the channel (**a** and **b**) with a Kuhl receiving no water (**c**) just after the monsoon season in November 2024 (photos were captured by the author)

3.5 Institutional Control and Maintenance Challenges

Although some Kuhls are under the control of state institutions, a limited success in the maintenance of Kuhls was observed (Sharma et al. 2015). Baker (2001b) highlights several socio-economic challenges of maintenance of Kuhls of Kangra Valley. Baker (2001b) mentioned such delays are uncommon in the developing nations due to release of funds by the state institutions. Such issues are very common and a hurdle in managing the traditional irrigation system effectively.

3.6 Riwaj-I-Abpashi: The Book of Irrigation Customs

A detailed description of Kuhls, their locations and few water-sharing customs can be found in the "Riwaj-I-Abpashi" often referred as "book of irrigation customs" documented during colonial era. Riwaj-I-Abpashi (written in Urdu with elements of Persian) is perhaps the only historical book that preserved Kuhl's information. The book mentioned over 3000 major and minor Kuhl's in the Kangra Valley was compiled with support from colonial power. The intention of compilation was unclear however, Baker (2001a) highlighted that this was done to control the water system for revenue collection.

The role of Riwaj-I-Abpashi document is pivotal, as it maintained Kuhl's historical information, water sharing customs and is only comprehensive documentary record on Kuhl (Baker 2001a).

3.7 Kuhl's Operated Watermills (Gharats)

Gharats are traditional local grinding watermills found throughout Himachal Pradesh (see Chap. 2). They are operated by using the kinetic energy of water from Kuhls or small streams and widely used earlier when there was no electricity. The watermill has several advantages, as it saves energy, generates no carbon, and is environmentally friendly and runs without electricity. Also, the generated flour is more nutritious, and healthier than that of electrical driven mills. This also reflects how nutritional value is declining with time in our food.

3.8 Decline of Gharats

Earlier, there was at least one gharat (traditional watermill) in every village. However, due to the intermitted water flows in the Kuhls, numerous gharats in the valley are now defunct. Some gharats currently operate seasonally, meaning they operate when the Kuhl receives sufficient water during the monsoon.

Despite many environmental and nutritious benefits, gharats are losing their identity reflecting an essence of losing traditional knowledge. Nowadays, people prefer electric mills over gharat[7], which require less energy to operate and generate more flour and economic return. Hence, gharat owners are also unwilling to continue, as watermills provide less return and no longer sustainable.

3.9 Current Challenges in Kuhl Management

Currently, Kuhls is a community and state institutions collaborative affair (Fischer 2017). However, the responsibilities of communities and institutions are poorly defined. This gap is clearly evident now in the Kuhls irrigation management system. While some of the Kuhls have been taken over by state institutions, others are still managed by communities thus, often led to conflicts over water management. Baker (1996) highlighted such gaps and the aftermath effects on irrigation. Moreover, we

---

[7] Interestingly, gharats flour has recently become available for purchase on grocery websites. Unfortunately, they are very expensive due to the high commissions charged by online platforms.

must agree that changes in irrigation patterns are inevitable and often tends to reflect the institutional interests, which lead to conflicts in a collaborative management effort.

Recent climate change and the construction of micro-hydropower plants in the upper reaches have posed significant challenges to the traditional Kuhl systems. As a result, Kuhls now received less water than usual (Ashraf and Akbar 2020). This directly affects numerous farmers who have historically utilized the Kuhl water for enhancing agricultural productivity during prolonged dry periods. For instance, high-return crops like rice can no longer be cultivated due to persistent water scarcity in many regions of the Kangra Valley (Sharma et al. 2015).

Reduced water flow in the Kuhl also affects gharat owners, who rely on water flow to operate their watermills. Although, irrigation through Kuhl water also has many challenges as the water sourced from snow/glacier melt is much colder than the groundwater, making them unpreferable for the crop growth. Currently, most of these earthen irrigation canals receive water only during the monsoon season, rendering local communities to grow year-round crops.

Uplifting water from nearby khads for irrigation is a costly affair. This requires substantial capital investment, which most farmers of the Kangra Valley cannot afford. Additionally, irrigation through uplifted water can only irrigate limited areas. In contrast, farmers in the adjacent Kullu District, are privileged of having a major river where water can be uplifted to cultivate orchards whose market value is higher. These high return crops make such investment more viable. Therefore, people are unwilling to invest in needed infrastructures in the Kangra Valley.

Although the entire agriculture field in the valley does not receive water from Kuhls. In several areas, the existing topography is not suitable to divert stream water, and thus, farmers only grow rainfed crops. For instance, Changar region[8] of Kangra generally referred to as unirrigated land because of the region rugged terrains and thus, only receives monsoonal rainfall, which further restricts Kuhl's development. A detailed account of the Kuhls is provided by Baker (2007).

3.10 Conclusion

This chapter underscores the crucial role of Kuhls in the Kangra Valley. Kuhls in Kangra are well-defined, gravity-fed irrigation channels, traditionally built to provide water for irrigation to boost agricultural productivity. The water from the streams is diverted into the small channels to flow water through gravity and used to provide water to farmland. Kuhls played a key role in sustaining agricultural productivity in the region.

Recent climate changes and human intervention, and inadequate policy frameworks, have resulted in reduced water flows in the Kuhls. Farmers also have the

---

[8] Changar regions are hilly tracts and rely entirely on rainfall for irrigation. Certain parts of the Kangra Valley fall in this region.

essence of climate change and thus, many of them have now shifted to rainfed culti-vated crops. This means less profitability, fuelling insecurity, and further perpetuates conflicts both within communities and families.

Presently, the Kuhl system is fading and losing its traditional heritage. Shifting the crops is now clearly evident and a clear sign of climate aftershocks. The gravity-driven channels once designed to nourish thousands of hectares of land area are now on the verge of dying. Although Kuhls were once the backbone of farming in the valley, as they provide support to grow multiple crops annually thereby making farming profitable.

Kuhls reflect how community wisdom, need, and willingness can dominate over highly qualified institutions to manage their water system for hundreds of years. The historical heritage system supported millions of people's livelihoods. Enhancing agriculture profitability could engage more people in the business. This in the long term generate employment and reduces migration to urban areas.

Kuhls are one of the finest examples of nature-based meaningful solutions for addressing water and food scarcity. Globally, limited examples of community-managed water resources are known, in which Kangra Valley is one. The unwritten and undefined rules are very well structured and managed by the communities in the valley despite many challenges. Kuhl plays a pivotal role in making profitable farming in hills, prospering businesses, and revitalizing agriculture. Kuhls are one of the best examples of community-managed water distribution systems. Low water flow in the Kuhls is a sign of great loss of traditional knowledge. Furthermore, increasing institutional intervention, communities often show less interest in Kuhl's management.

3.11 Recommendations

The conflicts over Kuhls should be resolved through dialogue and negotiation with the state institutions. Kuhls, proven as one of the best irrigation systems in the Western Himalayas, should be revived for water and food security in the valley and are important under the present rapid climate change conditions.

Although hydropower plants have their own advantages, the release of appropriate water in the Kuhls is utmost required to boost agricultural productivity. Furthermore, farmers should get their water rights by providing sufficient water in their Kuhls. Reviving Kuhls can offer a long-term sustainable solution for the Kangra Valley dwellers who are engaged in agriculture for their livelihood.

References

Ashraf A, Ahmad I (2021) Prospects of cryosphere-fed Kuhl irrigation system nurturing high mountain agriculture under changing climate in the Upper Indus Basin. Sci Tot Environ 788. https://doi.org/10.1016/J.SCITOTENV.2021.147752

Ashraf A, Akbar G (2020) Addressing climate change risks influencing cryosphere-fed Kuhl irrigation system in the Upper Indus Basin of Pakistan. Int J Environ 9(2):184–203. https://doi.org/10.3126/IJE.V9I2.32700

Ashraf A, Batool A (2019) Evaluation of glacial resource potential for sustaining Kuhl irrigation system under changing climate in the Himalayan region. J Mt Sci 16(5):1150–1159. https://doi.org/10.1007/S11629-018-5077-0

Baker M (1996) Changing contexts, steady flows: patterns of institutional change within the communal irrigation systems (Kuhls) of Kangra valley, Himachal Pradesh, India. Himalaya, J Asso Nepal Himalayan Stud 16(1). https://digitalcommons.macalester.edu/himalaya/vol16/iss1/6

Baker JM (2001a) Communities, networks and the state: continuity and change among the Kuhl irrigation systems of Western Himalaya. https://hdl.handle.net/10535/4563

Baker M (2001b) The politics of knowledge: the case of British colonial codification of "customary" irrigation practices in Kangra, India. Himalaya, J Asso Nepal Himalayan Stud 21(2). https://digitalcommons.macalester.edu/himalaya/vol21/iss2/7

Baker JM (2007) The Kuhls of Kangra: community-managed Irrigation in the Western Himalaya. University of Washington Press (Verlag). ISBN: 978-0-295-98764-4.

Coward E (1990) Property rights and network order: the case of irrigation works in the Western Himalayas. Human Org 49(1):78–88. https://doi.org/10.17730/HUMO.49.1.G53435062N67K2G2

Fischer HW (2017) Harnessing the state: social transformation, infrastructural development, and the changing governance of water systems in the Kangra district of the Indian Himalayas. Ann Am Assoc Geogr 107(2):480–489. https://doi.org/10.1080/24694452.2016.1232614

Gaurav K, Sharma P, Sharma A (2025) Revisiting community-based traditional irrigation systems in India. Heliyon 11(1):e41684. https://doi.org/10.1016/J.HELIYON.2025.E41684

Gómez-Baggethun E (2022) Is there a future for indigenous and local knowledge? J Peasant Stud 49(6):1139–1157. https://doi.org/10.1080/03066150.2021.1926994

Hanjra MA, Qureshi ME (2010) Global water crisis and future food security in an era of climate change. Food Policy 35(5):365–377. https://doi.org/10.1016/J.FOODPOL.2010.05.006

Khaneiki ML (2020) Cultural dynamics of water in Iranian civilization. In: Cultural dynamics of water in Iranian civilization. Springer International Publishing. https://doi.org/10.1007/978-3-030-58900-4

Malapane OL, Chanza N, Musakwa W (2024) Transmission of indigenous knowledge systems under changing landscapes within the Vhavenda community, South Africa. Environ Sci Policy 161:103861. https://doi.org/10.1016/J.ENVSCI.2024.103861

Misra AK (2014) Climate change and challenges of water and food security. Int J Sustain Built Environ 3(1):153–165. https://doi.org/10.1016/J.IJSBE.2014.04.006

Pinto AM (2014) Traditional knowledge and water quality in Timor-Leste: climate change adaptation strategies used by local communities in Laco-Mesac and Ulmera villages [Master Thesis]. Ohio University

Sethi AS (2011) Gravity flow irrigation system (Kuhl) in Himachal Pradesh — a case of Palampur region. In: 3rd IDSAsr international seminar on water security and climate change: challenges and strategies, pp 119–123

Sharma AK, Sharma KD, Prakash B (2015) Death of Kuhl irrigation system of Kangra valley of Himachal Pradesh: institutional arrangements and technological options for revival. Indian J Agric Econ 70(3):350–364. https://doi.org/10.22004/AG.ECON.230213

Part III
The Role of Institutions, Challenges, and the Way Forward

Chapter 4
Role of Institutions in Kangra Valley Development

Abstract Institutions play a pivotal role in shaping the lives of people. Institutions are the supervisors who formulate policies, schemes, allocate funds, ensure smooth functioning and achieving timely targets of schemes and bring innovations that can significantly change the lives of thousands of families. However, building a climate-resilient framework and promoting efficient water use practices is challenging due to limited infrastructures and uncertain climate change. This chapter underscores the institution's role in the development of Kangra Valley and ends up with suggestions to foster practical solutions. This study reveals institutions interfere often limit community participation. However, the study suggests a synergistic effort for sustainable development of the Kangra Valley of Himachal Pradesh.

Keywords Communities · Kangra Valley · Institutions · Himachal Pradesh · Western Himalayas

4.1 Introduction

Water resources are important for enriching communities, and bring prosperity and overall provide peace, security, and freedom. Institutions (the governing bodies) play a pivotal role in the sustainable management of natural resources of an area. By formulating good schemes, institutions can better manage natural resources, boost agricultural productivity, transform agriculture, and lower the rising prices. Therefore, the institution's role is crucial for developing feasible and sustainable solutions to better flourish human lives.

Balfour (2024) highlights the role of key actors such as politicians, clan leaders, and stakeholders in achieving sustainable water management. These players play a pivotal role in sustainable water management and their decisions could have large impacts on the indigenous people (Balfour 2024). Good decisions certainly improve farmer's livelihoods and resilience. This can be achieved by introducing innovation in agriculture so that farmers can produce enough agricultural goods and to ensure that this generation and also the future generations have access of sufficient water and

S. Ali, *Traditional Water Conservation Community-Managed Structures and Their Role in Valley Dwellers' Livelihoods*, SpringerBriefs in Water Science and Technology, https://doi.org/10.1007/978-3-032-04637-6_4

food. Furthermore, the institution's plays a pivotal role in enduring life in conflict-affected and resource scarce communities. This is also true for the Kangra Valley of Himachal Pradesh.

Water and food scarcity exacerbate the risk of conflicts. Hence, in Himachal Pradesh, state institutions are making every effort to develop sustainable water and food security schemes. For instance, emphasis is given on producing high quality organic fertilizers, developing seeds naturally suited to local conditions, adopting smart agriculture, encouraging mixed farming, and certified seeds to promote climate-resilient agriculture. The state institutions have increasingly subsidized organic farming to foster natural farming using negligible harmful chemicals.[1] Institutions are also seeking international aid to partially support farmers to enhance agricultural productivity and support businesses. For e.g., Himachal Pradesh Subtropical Horticulture, Irrigation & Value Addition[2] (HPSHIVA) project was aim to ease the agricultural business by introducing climate resilient farming, adopting crop diversification, and providing appropriate market to the farmers. The scheme also provides subsidies for fencing to protect crops from wild animals.

Although numerous schemes were designed in a timely manner, and there is no shortage of policies in the country, a considerable gap persists between policies and practice. This reflects a clear gap in developing a policy that could be fully implemented. Thus, there is a significant gap in converting policies into practices. Major constraints are inadequate monitoring of resources, limited infrastructure, lack of funding, unscientific farming practices, lack of quality research, delay in taking decisive actions, and weak policies towards conserving the environment. Addressing these challenges is essential for bridging the gap between policies and practices. Thus, this study documents key points to achieve the long-term sustainable goals in the Kangra Valley of Himachal Pradesh.

4.2   Suggestions for Achieving Long-Term Sustainable Goals in the Kangra Valley

I here listed key recommendations that could be considered by the institutions to foster better management of the resources in the Kangra Valley and to work towards the vision of prosperity, resilience, and sustainability of the valley dwellers (Fig. 4.1).

---

[1] Himachal Headlines news https://www.himachalheadlines.com/news/himachal-takes-major-step-to-boost-natural-farming-corn-purchased-from-400-plus-farmers/ (Accessed on 21–10–2024).

[2] https://hpshiva.hp.gov.in/cms//en/home/.

Fig. 4.1 Key points to foster sustainable resource management in the Kangra Valley

4.2.1 Fostering Traditional Indigenous Knowledge

Traditional indigenous knowledge has supported local communities for generations. Local traditional knowledge often undermines by the institutions during policy
formulation. Local wisdom supports communities and offers practical nature-based
solutions. The traditional knowledge not only cultural but also technical and plays
a pivotal role in sustainable management of natural resources and climate resilience
(Mekonnen et al. 2021). Thus, the institutions should promote and incorporate
traditional knowledge to empower communities and environmental sustainability.

4.2.2 Establishing Local Collaboration

Communities' participation has a major role in the development and management
of natural resources (Shunglu et al. 2022). Local governments alone are unlikely
successful in achieving the long-term sustainable goals related to water and food
security without the active participation of local communities. Farmers and locals
have deep knowledge of their farms better than anyone and often passed down through

generations. Therefore, community's role cannot be neglected because they know practical solutions and their traditional knowledge can be very helpful in providing sustainable practical solutions.

4.2.3   Strengthening Local Businesses

Local businesses support communities and crucial for economic resilience. Institutions should provide supports to businesses to sustain local livelihoods. For instance, livestock grazing has several environmental benefits. Livestock grazing can improve soil health by enhancing soil fertility, reduce soil erosion, improve the carbon storage capacity in the soil, and support biodiversity (Oerly et al. 2022). Thus, strengthening such local businesses can contribute to rural prosperity and ecological conservation in the Kangra Valley.

4.2.4   Empowering Women

Women are often deprived and under-examined in developing countries, despite most vulnerable to any environmental impacts. However, without them, the sustainable development of the valley cannot be achieved. Even though, in many schemes, women are overlooked in many development schemes. This perception towards women should be changed. Women must be encouraged, included, promoted, and empowered by actively involving them in every scheme by ensuring that their contributions should be valued. Their role is crucial in conservation of natural resources of the Kangra Valley. Although some local committees of women were established, but their role is weakly defined and fragmented. Empowering women is important for the long-term sustainability of the region.

4.2.5   Reviving Natural Resources

All natural resources should be revived to address growing challenges related to water and food scarcity in the Kangra Valley. In recent years, water resources are gradually declining due to reduced frequency of rainy days and longer and more frequent dry spells, resulting in water shortage in khads, lakes, and springs. These water sources should be conserved and rejuvenated to build a resilient future for the region. Thus, there is a need to promote and strengthen the natural water conservation practices.

4.2.6 Preparedness for Climate Change

Frequent extreme weather events like prolonged droughts and floods, are continuously threatening agricultural productivity and damaging natural ecosystems. In the Kangra Valley where most of the dwellers are depends on farming for their livelihoods, such extreme events are often devastating. Therefore, preparations towards any climate uncertainties are important to save from heavy losses and conserve the natural resources.

Institutions must provide timely training and capacity-building programs to prepare communities for natural disasters. These includes climate resilient farming, awareness campaigns, and practical adaptive practices. Strengthening local preparedness will empower local communities to respond more efficiently to any environmental calamities. Additionally, the digitization of natural resources is crucial for establishing early warning signals to reduce damages from climate extremes.

4.2.7 Digitization of Natural Resources to Strengthen Disaster Preparedness

All natural resources should be digitized to alert against any natural disaster and better prepare for climate extremes. Digitization is also imperative for guiding farmers about the weather patterns, enabling them in adopting crop selection, timely cultivation, and informed decisions.

Additionally, maintaining the digital records of water resources can help in developing effective conservation strategies. These measures are important for conserving naturals resources from natural disasters and human's encroachment. Although digital innovation in India is still in its nascent stage, expending its applicability in the agricultural sectors is essential for building climate resilient and sustainable resource management.

4.3 Other Suggestions for the Development of the Kangra Valley

The role of institutions is important in addressing the urgent needs of vulnerable communities and in guiding them towards sustainable and resilient livelihoods. By introducing practical based subsidy schemes to the farmers, agricultural productivity of the farmers will be enhanced. It must be clearly designed in such a way that the intended farmers are benefited.

Although beneficiaries getting more benefits from the schemes is remains a persistent challenge. Kafle and Balasubramanya (2023) argued that although more emphasis is given on the supply side, the demand side is always limited studied during the formulation of policies. Hence, this led to significant gap between demand and supply side interventions. Addressing these challenges requires holistic approach in planning and implementing policies for the Kangra Valley.

Institutions should also focus on developing other water harvesting structures that are defunct now. Water conservation structures such as khatris, which are less common in Kangra District should be developed. Khatris are again an indigenous example of water conservation structures that reflect indigenous traditional knowledge of the people, which is completely lacking in the Kangra Valley. Furthermore, boweris should be revived where they are defunct due to human intervention and tap water supply. Check dams/recharge trenches on perennial khads should be constructed and promoted, which could significantly recharge the groundwater and dying springs (Balooni et al. 2008). All these water conservation structures are century old and the wisdom communities gained to build them is through experience, need, and knowledge.

My literature review across major well-known scientific engine sites such as Google Scholar, ScienceDirect and random searches in Google reveals limited studies focusing on the Kangra Valley. The lack of research highlights the urgent need for more quality research, which could be easily access by everyone. This will deepen our understanding of natural resources of the Kangra Valley and guide more effective development strategies.

4.4   Challenges in Developing a Sustainable Kangra Valley

There are numerous challenges in achieving sustainable development in the Kangra Valley, facing several challenges related to food and water scarcity. One such concern is the growing groundwater demand in the valley in response to surface water scarcity. To meet this demand, residents and developers are increasingly drilling deep wells to extract groundwater. For example, newly constructed residential flats in the valley often require continuous water supply and often meet through groundwater. This growing demand of the groundwater can pose serious environmental concerns. Since, the groundwater in the region is already limited, over extraction could lead to rapid depletion. This will increase the vulnerability of drought which are an important source of recharge to springs. Thus, large dependency on the groundwater is an essence of future water insecurity and require urgent institutional attention to mitigate this issue with suitable interventions.

Sometimes, the adoption of new techniques overrides traditional ones, and exacerbates existing challenges. For instance, many Kuhls have been cemented with an intention to reduce water loss through infiltration. Over time, state institutions of Himachal have taken over most of the Kuhls, leading to a decline from the community participation.

Furthermore, the construction of micro-hydropower plants in the upper reaches along with efforts in providing water for drinking from major khads to every household, by and large, reduces water flow in Kuhls. This realizes how water in Kuhls which once flowed throughout the year, is now arrives intermittently.

Currently, water supply schemes sourced from khads also face significant challenges during the monsoon season. Heavy rain during monsoon often carries huge silts with them, and clog water channels, thereby disrupting water supply for the extended periods. This situation is further fuelled by irregular maintenance and inconsistent cleaning practices, rendering them from regular water supply to the households.

Another major challenge in the Kangra Valley is the diverse geography of the valley. For instance, Kuhls cannot be developed in many regions due to unsuitable terrain, restrict areas to grow only rainfed crops. One such example is the *Changar region* of Kangra, as discussed in Chap. 3.

Policy implementation often suffers from change of state and central government. Maintaining good coordination between state and central government is always challenging if both are from different political parties. This further hurdle in implementing policies, and the outcomes are often devastating.

Recently, in a significant move, institutions have initiated the rejuvenation of gharats (watermills) under the Mahatma Gandhi National Rural Employment Guarantee Act (MNREGA) scheme.[3] This was aimed to restore lost gharats in the valley to preserve the indigenous knowledge and promote local businesses. However, the availability of water for running the gharat is remain uncertain, as water flows in Kuhls is often intermittent. Furthermore, the profitability of these enterprises is also low (see Chap. 3).

Although the primary focus of farmers is always on enhancing agricultural productivity enhancement rather than water conservation. Therefore, the institution's role is pivotal in shifting the mindset of farmers and promoting effective water management practices. However, this is often challenging in the present climate uncertainties.

4.5 Conclusion

Institutions play a crucial role in the sustainable development of the Kangra Valley. Although, climate change has largely impacted the region's natural resources, this vulnerability is further intensified by weak institutions. Even though, concerned authorities are making efforts to preserve the resources and enhance agricultural productivity, however, these efforts are inadequate and often fragmented.

Additionally, greater emphasis should be given on effective water conservation measures. Institution's coordination with the local communities is warranted for developing meaningful, practical nature-based solutions. However, somewhere it reflects that institutions want more power in decision making.

---

[3] https://www.youtube.com/watch?v=jG19JfPdewk (accessed on 20–11–2024).

Policy implementation is often delayed due to the frequent transfer of supervisors, limited infrastructure, lack of willingness, degree of attention paid, and other local reasons. Institution's long and unnecessary procedure to avoid corruption charges, significantly impacted the implementation of the schemes. Thus, delay in decision making in a government scheme is inevitable. More often, institution's frameworks are fragmented and lack effective community coordination.

People living in the mountains have a high level of trust with the institutions. This is because of easy access to the concerned institutions, reflecting the welcoming culture that still exists in the mountains. This facilitates direct communication to the concerned officials, which often lacks in other regions.

Key challenges such as climate change, land use alteration, land degradation, etc., poses significant threat to valley dweller's livelihoods. This is further compounded by unpredictable climate patterns makes the future more complicated and uncertain. Climate uncertainties are increasingly dangerous and unmanageable over time. Marginal and tenant farmers, women, and children are highly vulnerable to climate shocks. Hence, the current situation calls for an urgent policy intervention.

Climate change is pervasive, and thus there is a need to change ourselves with the present climatic conditions. This is further exacerbated by limited natural resources and a rapid rise in population. Nature-based solutions in the present unpredictable climate conditions offers sustainable solution. This requires a holistic approach by bridging knowledge gap between policies formulation and practical implementation. Thus, policies should focus on deploying proven strategies supported by clear rationales.

Sustainable development is a shared responsibility, and thus fostering synergies between local traditional knowledge and global expertise can advance sustainable development effectively. Such efforts will protect water resources, generate more employment, reduce poverty, and inequality, and promote resilience, ultimately improving Kangra Valley dwellers livelihoods. However, we must agree that formulating perfect policies is unrealistic and unattainable.

To conclude, the integration of local traditional knowledge and global expertise can effectively address many urgent challenges related to water and food scarcity, thereby paving the way for the sustainable development of the Kangra Valley.

References

Balfour N (2024) Groundwater management in the horn of Africa: conflict, scarcity, and hybrid governance (English). World Bank Group. http://www.worldbank.org/

Balooni K, Kalro AH, Kamalamma AG (2008) Community initiatives in building and managing temporary check-dams across seasonal streams for water harvesting in South India. Agric Water Manag 95(12):1314–1322. https://doi.org/10.1016/J.AGWAT.2008.06.012

Kafle K, Balasubramanya S (2023) Can perceptions of reduction in physical water availability affect irrigation behaviour? Evidence from Jordon. Clim Dev 15(5):353–365. https://doi.org/10.1080/17565529.2022.2087587

Mekonnen Z, Kidemu M, Abebe H, Semere M, Gebreyesus M, Worku A, Tesfaye M, Chernet A (2021) Traditional knowledge and institutions for sustainable climate change adaptation in Ethiopia. Curr Res Environ Sustain 3:100080. https://doi.org/10.1016/J.CRSUST.2021.100080

Oerly A, Johnson M, Soule J (2022) Economic, social, and environmental impacts of cattle on grazing land ecosystems. Rangelands 44(2):148–156. https://doi.org/10.1016/J.RALA.2021.12.003

Shunglu R, Köpke S, Kanoi L, Nissanka TS, Withanachchi CR, Gamage DU, Dissanayake HR, Kibaroglu A, Ünver O, Withanachchi SS (2022) Barriers in participative water governance: a critical analysis of community development approaches. Water 14(5):762. https://doi.org/10.3390/W14050762

Chapter 5
Towards Sustainable Development in the Kangra Valley: Challenges and Prospects

Abstract This chapter explores the challenges and prospects for sustainable development in the Kangra Valley, emphasizing the urgency to address emerging water and food crises. It highlights how climate change introduces great uncertainties, which are further exacerbated by anthropogenic activities, and fragmented policies. These factors collectively hinder the effective management of the valley's water and food resources. The chapter concludes with strategic recommendations to empower local communities and improve the livelihoods of valley dwellers.

Keywords Agriculture · Communities · Kangra Valley · Himachal Pradesh · Traditional indigenous knowledge

5.1 General Opinion

"If the indigenous knowledge is so powerful, then why my farm is poor"? This question posed by a farmer to the author Briggs (2005) is quite logical and draw careful attentions and compels us where we lack the ability to provide meaningful results even having numerous traditional knowledge which took many years to develop. The answer to this question raised by a farmer to a researcher with no formal education degree is significant. According to him, if knowledge cannot improve livelihood, then it is a waste. In my opinion, the farmer's point is logical and if such questions cannot be addressed by researchers and institutions with feasible results, it's an essence of failure. Hence, it is our responsibility to convince the farmers with practical, feasible solutions, not by convincing theories.

My field survey reveals that the most farmers and even elders, not directly involved in agriculture, have deep and insightful knowledge about water.[1] Their practical knowledge and critical thinking are impressive, realising that local people understand their farms better than the institutions and know experienced and practical solutions.

---

[1] The mode of communication was Kangri/Gaddi language in most of the time (and Hindi quite often) so that responders could better express themselves. Also, responders were not informed about the intention of asking the question to express the real situation.

  57

S. Ali, *Traditional Water Conservation Community-Managed Structures and Their Role in Valley Dwellers' Livelihoods*, SpringerBriefs in Water Science and Technology, https://doi.org/10.1007/978-3-032-04637-6_5

The traditional knowledge passed down through generations has evolved through many modifications, continuous discussions, and years to develop, which often seems very simple and easy to us today. Their wisdom supported them to survive and thrive for decades. Therefore, institutions should acknowledge them and integrate their practices with modern techniques when designing any water or food-related schemes.

My interaction with numerous valley dwellers revealed that people are not opposed with the developments and increasing tourist's traffic in the valley however, they are deeply concerned to protect their natural resources. Furthermore, the traditional resources were always a survival source for valley dwellers, and they have an essence that the loss of these resources creates a sense of insecurity. There is awareness that these natural assets are irreversible, and at the same time the transformation is inevitable.

The role of institutions is pivotal in managing the natural resources. I asserted in Chap. 4 that institutions must work closely with local communities to develop effective water management strategies and bring innovations in agriculture. I argued that cutting-edge advanced modern technologies should be adopted without neglecting the traditional one. For sustainable development, it is essential that the valley becomes self-sufficient in agricultural production.

All traditional water systems in the Himalayas are "constructed by experience". The experience came from the demand of water for livelihood. Whether it is to meet the demand for drinking or domestic use or for irrigation, these systems are direct response to challenges. The traditional wisdom passed down through generations are powerful example of how communities used nature-based sustainable solutions in the past, supported them for centuries. As nature-based solutions proved their value in building resilient communities, it is essential that institutions acknowledge and incorporate traditional knowledge for the greater benefit of local communities.

5.2   Key Points for the Sustainable Development of the Kangra Valley

This section highlights several key recommendations that can be taken into account for developing a prosperous and sustainable Kangra Valley.

5.2.1   Clear Institutional Responsibilities and Timely Performance Monitoring

A crucial issue in the resource management is the lack of clearly defined responsibilities among institutions. This usually creates chaos and often leads to confusion, inefficiency, and ultimately mismanagement of resources. Establishing clear guidelines and frameworks for institutional accountability is essential. Timely evaluations and performance assessments can ensure that the tasks are executed effectively. The optimum use of the resources can only be had through well-defined responsibilities and coordination.

5.2.2   Strong Commitment to Institutional Responsibilities

A strong commitment to assigned responsibilities is essential for institutional effectiveness. A strong commitment towards completion of any task can address many inefficiencies proactively. Lack of strong commitment often led to delay, mismanagement and always devastating. Strong institutional commitment is the key for sustainable solutions in the Kangra Valley.

5.2.3   Addressing and Accepting Shortcomings and Inefficiencies

Recognizing and addressing shortcomings and inefficiencies is a crucial aspect of institutional growth and development. Often institutions fail to acknowledge their inefficiencies, rendering further progress and weaken public trust. Accepting and addressing limitations and taking timely actions are essential for achieving sustainable development goals. Hence, institutions should regularly self-access their incapabilities and commit for self-improvements in order to effectively serve the growing needs of the Kangra Valley.

5.2.4   Collaboration with the Local Communities

Local communities always have deep and experimental knowledge on their natural resources. Communities living for many years in a specific area have developed various nature-based solutions to support their survival and agricultural livelihoods. The traditional knowledge invented by the Kangra Valley dwellers took many years to develop and result of their continuous observation and innovation. Institutions must recognise the value to indigenous wisdom and include in planning and implementation to strengthen schemes related to agricultural productivity and water conservation. Collaborating closely with local communities is essential to strengthens the relevance of development interventions and for long-term sustainability.

5.2.5 Expanding the Role of Funding Organizations Beyond Financial Support

International funding organizations should not limit themselves to providing financial support. In order to enhance the sustainable development of the valley, these organizations should also provide appropriate infrastructure, skilled manpower, and technical expertise. Their involvement in building local capacity, through training and knowledge sharing with local communities can enhance the effectiveness of development efforts. Their extensive reach and access to highly qualified professionals, funding organizations can empower local communities with the tools and skills needed for sustainable development in the Kangra Valley.

5.2.6 Frequent Advanced Technology Training to Bridge Knowledge Gaps

Institutions and organisations should conduct regular workshops to transfer the knowledge to the local people to enhance their agricultural productivity. The training should not be only limited to the institutions but reach to grassroots level. This requires developing proper and feasible framework for providing timely, accessible training to local communities.

Encouraging the participation of women and youth to contribute actively to agricultural development will empower them. Building strong citizen science initiatives is essential to foster community engagement.

5.2.7 Strengthening Efforts to Preserve Water-Conservation Structures

Climate change poses an increasing threat to water conservation structures, necessitating their preservation for managing water resources and sustaining both agricultural and domestic needs. To ensure year-round water availability, there is a need to harvest rainwater and store the excess drain water during periods of abundance. Constructing check dams on khads to recharge the groundwater and springs in the downstream is highly recommended. Additionally, traditional water system such as boweris should be rejuvenated throughout the Kangra Valley. Further, the regular maintenance of existing boweris is crucial to ensure these systems remain functional and continue to support local communities.

5.3 Challenges in Addressing Sustainable Development of the Kangra Valley

As mentioned earlier, climate change brings large uncertainties, challenging sustainable development efforts. Hence, climate-resilient agriculture schemes have been introduced and increasingly promoted in the Kangra Valley. Despite their intent, these schemes bring their own challenges, particularly regarding their feasibility, economic profitability, and social acceptance.

For instance, rice crops offer higher market prices than wheat and attract farmers. Although paddy cultivation requires nearly three times more water than wheat, raising significant concern about environmental sustainability. Rajput et al. (2024) have suggested that crop shifting can save substantial water in plain areas. Nevertheless, the economic returns from the alternative crops often do not match with paddy, thus despite numerous efforts, groundwater levels continue to decline. By addressing farmer's income security and highlighting the benefits of other water efficient cereals (such as millets) could help in mitigating this growing water concern while supporting farmers livelihoods.

Even though Kangra receives higher rainfall in its upper reaches and has numerous perennial streams, is now experiencing increasing water scarcity. This requires the urgent need for more efficient water management practices. In this context, the valley could learn from arid countries that optimally use their limited water resources (Marin et al. 2017). However, it is important to recognize that technologies utilized in water-scarce countries like Israel cannot be exactly replicated, nor they show likely the same results in India. This is due to numerous local factors such as available infrastructures, funding availability, institutional capacity, public perception, willingness of institutions to maintain and implement such schemes. Most importantly, technologies developed in Israel is driven by prolonged water crisis, which often leads to such discoveries. Therefore, Israel serves as a perfect example globally (Marin et al. 2017). Rather, sustainable water management in Kangra will require innovations that must be rooted in local environmental, social, and institutional context.

Drip irrigation faces many challenges in India, as highlighted by Alam et al. (2024). In Himachal Pradesh, drip irrigation has been promoted to enhance crop productivity and adopted in limited valleys. Although its application in the Kangra Valley remains challenging due to hydrological and geographical constraints.

Unlike other regions, Kangra lacks any major river and the perennial streams often carry less water during the dry season. This makes drip irrigation challenging in the valley. In contrast, in the adjacent Kullu Valley, water is been lifted from the Beas River, making drip irrigation practical.

The absence of any major river in the Kangra Valley makes lifting water from streams unreliable, as these streams often carry less water during dry periods. However, in the Palampur area, an attempt was made earlier to lift water from a major khad (Singh 1982). Although such effort faces significant barriers as lifting requires high energy, infrastructure, and often costly, making it less viable solution for most parts of the Kangra Valley.

Heavy loss of potable water in the valley disrupts the water supply and is typically higher in the mountainous regions compared to the plains. This is further compounded by lack of continuous monitoring and inadequate infrastructure, which is responsible for the poor water supply mismanagement. Therefore, natural water sources remain a reliable and important source for communities throughout Himachal Pradesh.

Climate change often brings large uncertainties, disrupt farmers livelihood with farmers with limited land/marginal farmers, and tenant farmers and underprivileged families are often more vulnerable to climate uncertainties (Jamshidi et al. 2019). In some cases, families are rented out their farms, and thus reducing the profitability by half.

Despite these challenges, farmers have developed waste land into productive agricultural land through their wisdom. For instance, terrace and contour farming techniques help in reducing soil erosion and transform wasteland to productive farmland. However, such efforts are currently limited in scale and contribute only marginally to overall agricultural productivity.

Furthermore, Kuhl irrigation is the best example of a traditional community-managed water system in the Western Himalayas, is now losing community participation. This decline is closely linked to the substantial drop in agricultural productivity due to reduced water availability in the Kuhls, also resulting in declining community participation. Community participation offers significant benefits as they regularly maintained Kuhl throughout the year for smooth water flow.

To ensure the continuous flow of water in the channels (Kuhls), especially during the peak monsoon season when water velocity and silt content are high. The community appointed Kohlis (Kuhl's supervisor) whose primary role was to maintain the irrigation channels and ensure uninterrupted water flow in the Kuhls throughout the year (see Chap. 3). Kohlis also served as skilled negotiators to resolve any water-sharing conflict. However, their role is gradually fading due to increased state intervention. This could further exacerbate insecurity in water availability and distribution to farms.

With increasing institutional intervention, there is now a noticeable decline in community participation, despite their importance in local agriculture. Kuhl's offers numerous benefits as the farmers have the privilege to grow multiple crops round the year. This enhances socio economic conditions of the farmers and supports local businesses. Otherwise, farmers are limited to only rainfed single crop, which reduces profitability and negatively impact economic conditions of the farmers, and their families.

Currently, there is a lack of community engagement in water conservation efforts, a trend often observed due to institutional intervention. This raises a primary question, if the agriculture sector continues to decline, how will the valley meet its food demands? In the long term, this will accelerate food insecurity and thereby enhances food prices and increased additional stress on rural economies.

Although the institutions are making considerable efforts to promote agriculture, however a significant decline could be observed unless the existing frameworks are revisited and feasible community inclusive schemes were not formulated. Without such proactive steps, Kangra Valley may face a severe water crisis challenges and long-term socio-economic instability.

Kangra Valley is distinguished by its rich indigenous culture in the Global South. Currently, its traditional water heritage in the valley is dwindling, impacting not only agricultural productivity but also other local businesses. For instance, watermills (*Gharats*) are defunct due to lower water flows in the Kuhls. According to valley dwellers *"if there is no water, then nothing grows in the valley"*. Profitability remains low when farming relies solely on rainfed crops while high agricultural productivity incentivizes farmers to continue cultivation. Thus, addressing future challenges related to food and water security is crucial for the Kangra Valley sustainable development.

Furthermore, groundwater resources in the Kangra Valley are limited, particularly in the mountainous areas. Installing tube wells is costly, and unaffordable for marginal farmers. Moreover, groundwater occurs at a deeper depth, except near the khads, where groundwater is relatively shallower due to continuous recharge. These are the only regions in the valley where most of the paddy cultivation is confined due to sufficient availability of water to support such water-intensive crops.

There is a growing drinking water quality concern in the Kangra Valley. For instance, Thakur and Panda (2012) reported that the drinking water supply in the valley is contaminated with both organic, and inorganic pollutants. Dev and Bali (2019) also reported that the groundwater in majority of the area is unsuitable for drinking and irrigation purposes. Water resources in other neighbouring districts have also found to be contaminated by various pollutants (Thakur et al. 2018). These findings highlight urgent need for continuous monitoring of drinking wells to prevent any waterborne diseases.

However, despite these risks, many valley dwellers consume water without prior treatment due to traditional cultural practices. This tradition is continuous from the earlier generations who used to drink same water. In many instances, both people and livestock drink from the same water sources, a practice seem concerning. However, this perception is changing with growing awareness and education recognizing the importance of safe drinking water on health.

Kangra is largely influenced by traditional knowledge history on how communities had managed their natural resources offering valuable lesson to the globe. Historically, the communities have well managed their natural resources without any central interventions, relying solely on collective action and participatory behaviour of the communities. This reflects how developing countries have surprised the developed world in managing their natural resources by local effective way. The valley exemplifies how local knowledge and community-driven practices can be more effective than formal institutional policies in managing the local natural resources. However, these practices must be continuously supported and strengthened.

5.4 The Way Forward for Kangra Valley

This section outlines the critical steps to safeguard local resources and build more water-resilient and sustainable valley. Water availability for both irrigation and domestic use is currently dwindling at an alarming rate across the globe and the Kangra Valley is no exception of facing such challenges. Addressing these challenges requires context specific strategies. Accepting the shortcomings from the past experiences and developing effective long-term solutions is the first step towards achieving water-resilient valley.

Adapting agricultural practices in response to climate change is essential. Farmers are also aware of climate changes and their consequences on water and agricultural productivity. Thus, a systematic approach is needed to support valley dwellers.

Historically, indigenous and traditional knowledge have played a pivotal role in managing the natural resources. Thus, the local wisdom should be valued because these practices have proved to be successful in managing the local resources.

Comparing Kangra Valley to other valleys may lead to misleading conclusions. Each region is characterised by different scenario that requires distinct strategies rather than a one-size-fits-for-all approach. Thus, water management schemes must be designed by incorporating local prevailing conditions.

Ensuring the valley's self-sufficiency towards water and food is essential. A decline in agricultural productivity could directly impact employment in the valley, leading to increased economic insecurity, driving a potential rise in conflict and crime.

Like any other developing nation, women in the Kangra Valley remain underrepresented in agriculture and other sectors. Women led families are only engaged in the businesses. Cultural practices and norms remain a barrier in the developing nations especially in rural areas of India. Therefore, efforts must be made to involve women participation, ensuring gender-inclusive growth and empowerment.

Nature-based or results-driven meaningful solutions offers effective, feasible, and acceptable path forward. Institutions should promote such interventions that integrate traditional wisdom with modern innovations. Advanced technologies that help farmers better understand climate uncertainties and enhance resilience to climate change should also be promoted. Adopting climate smart agricultural practices will improve productivity and reduces costs and essential for well-being of largely agrarian population of Kangra Valley.

However, numerous challenges remain. For instance, gathering accurate and advanced water-related data is challenging. This requires adequate funding, infrastructure, and skilled personnel. Although state and central institutions are making every effort to collect data, the lack of digital infrastructure makes accurate and timely data collection difficult. Digitizing water data is essential to enable informed and judicious decision-making.

One of the critical traditional irrigation systems in Kangra Valley is the Kuhl system, which provides insurance to agricultural productivity against monsoon weakness (see Chap. 3). However, several Kuhls are under threat due to ongoing infrastructure development. For instance, Kuhls have been damaged during road widening projects. This is further proliferated with large infrastructure development. This reflects how communication gaps between institutions can have devastating results.

With the decline of the Kuhls, the role of Kohlis is also fading. These individuals possess strong knowledge in managing their water systems exhibits strong conflict resolution skills. Institutions should recognize and integrate Kohlis into water governance framework. Hence, they should be recognised as "water warriors" or appointed in roles similar to the "Bhujal Jankar[2]" (groundwater expert) in western states; Gujarat and Rajasthan (Chinnasamy et al. 2018).

Furthermore, Kuhls are on the verge of dying, showing how traditional knowledge is losing its heritage with time, which was once a boost for crop production and a reliable source for many years to the Kangra Valley farmers. The loss is threat to local agricultural sustainability.

Currently, Kuhl maintenance is undertaken by the state government under the Mahatma Gandhi National Rural Employment Guarantee Act (MGNREGA).[3] However, delays in fund disbursement often hinder maintenance of these irrigation channels. Thus, the release of funds for traditional water conservation structures should be accelerated without any delay.

Additionally, advanced techniques such as aquifer storage and recovery (ASR[4]) should be promoted, considering the valley's high monsoonal rainfall. However, in ancient times, rainwater was stored in khatris, unintentional yet effective storage systems (see Chap. 2) mainly seen in the adjacent Hamirpur District. These tanks were constructed for drinking purposes are now largely extinct due to the widespread tap water supply. Moreover, reviving such traditional structures could help store monsoonal rainwater and could be used during the dry periods.

Although Kangra Valley dwellers are identifying opportunities to enhance agricultural productivity for livelihood improvement, the adoption of new technologies remains a significant challenge. One of the key barriers is the results outcome in the long run from these innovations, which particularly affect underprivileged societies who are more susceptible to rumours. Similarly, climate adaptive irrigation techniques face similar issues with acceptance. The long gestation period for returns on investment and the difficulty in communicating clear cost–benefit outcomes make technologies like drip irrigation less appealing and feasible for many farmers in the Kangra Valley.

Proactive measures such as transforming agriculture through digital innovation, enhancing data collection, identifying correct environmental problems and promoting climate-smart agricultural technologies can drive significant reform in

---

[2] Local volunteers involved in groundwater management (https://www.marvi.org.in).

[3] Guaranteed income scheme passed in 2005 by the central government of India.

[4] ASR is the method to store the water during the wet season when plenty of water is available and recover in the dry seasons.

the agricultural productivity. For instance, the introduction of resilient, productive, and innovative crop varieties could play a critical role in increasing agricultural sustainability and improving yields under changing climatic conditions. Equally important is the emphasis on bilateral coordination among practitioners, researchers, policymakers, and institutions. Such collaboration ensures that interventions remain context-specific, scientifically grounded, and aligned with the real needs of farming communities.

Although not all signs are negative, encouraging progress have been made in several areas. For instance, the adoption of climate-resilient seed varieties has substantially improved crop production while reducing water usage. In another positive development, state institutions have initiated flood protection measures to rehabilitate and manage khads. Additionally, NGOs supported by both national and international agencies are actively engaged in the restoration of khads. The initiative also includes the rejuvenation of traditional water systems across the valley. However, the progress remains slow, limited, and often fragmented. These interventions need community-driven strategies for the long-term environmental sustainability.

New insights reveal a marked decline in interest among the younger generations in farming. The farmers assign many reasons for this. First many young people prefer the government jobs, which involve less uncertainties compared to farming. Second, agricultural profitability in the region has reached at its lowest points making it less profitable livelihood option. Third, with expanding families, lands were further subdivided, making it difficult to sustain a profitable farming on limited land.

Additionally, there is a cultural shift wherein young generations do not want to involve themselves in physically demanding agricultural labour. Given Kangra's growing tourist destination, many prefer to engaged in tourism sector. New trends to migrate abroad for better opportunities are quite common in the valley. These trends underscore the urgent need for policy and institutional interventions that make farming more appealing, viable, and rewarding for the next generation.

Institutions must recognize the critical role of young generations including women in the conservation of local natural resources. These groups often exhibit a strong commitment to environmental stewardship and should be involved in planning and decision-making processes to conserve local resources sustainably. Furthermore, the transfer of traditional indigenous knowledge to the younger generation is a shared community responsibility, with elders playing a particular crucial role within the communities. Elders play a vital role in motivating and guiding youth by passing down traditional wisdom. As younger individuals often model their behaviour after their elders, this intergenerational exchange is essential to preserve local traditional wisdom and ensuring long-term conservation efforts.

To encourage youth in environmental stewardship, the school education system should be redesigned to incorporate climate change issues. By embedding real-world environmental challenges into the curriculum, students can develop a deeper curiosity, problem solving mindset towards natural resource management. This early exposure will equip future generations with the knowledge and motivation to actively participate in conservation efforts as they progress through school and higher education. Therefore, curriculum development with a strong emphasis on sustainability and climate resilience will have remarkable results. These are a few small, yet crucial and impactful steps in fostering a culture of responsibility towards conserving our natural resources.

In many urban cities, younger generations often perceive indigenous knowledge as outdated or irrelevant in modern time. However, the youth of Kangra Valley know the value of traditional knowledge as narrated by their elders to them. My interaction with college students aged 18–25 surprised me, as they have strong awareness and appreciation for the local traditional knowledge systems, particularly in the context of water management and natural resource system. This awareness appears to stem from intergenerational learning, with elders actively sharing their experiences and values traditional wisdom. Strengthening this knowledge transfer can not only deepen youth engagement with natural resources but also motivate them to view farming as a meaningful and sustainable livelihood.

Additionally, the majority of the farmers in Kangra Valley are marginal, meaning they hold small farms. These limited resources constrain their ability to generate substantial profits from agriculture. Thus, many farmers have been compelled to abandon traditional farming practices and seek alternative livelihoods. This prompted valley dwellers to shift toward tourism-related businesses, driven by the valley's growing appeal as a tourist destination. While this transition offers new economic opportunities, it also raises serious concerns about the long-term viability of agriculture.

There is an African proverb that says *"If you want to go fast, go alone. If you want to go far, go together"*. This is also true for the agriculture sector. Synergy between local communities and institutions can address critical water and food security challenges and significantly contribute to the sustainable development of Kangra Valley.

Institutions should actively involve local communities in the formulation of water saving schemes to effectively address water scarcity issues. Somehow, we must think rationally to recognize that the responsibility for managing natural resources does not lie solely with institutions. While institutions play a critical role and continue to learn from both past and present experiences, effective water management requires a collaborative approach. The perception that only paid professionals or government bodies are responsible for environmental stewardship must be challenged. Sustainable natural resource management is a shared responsibility and every individual has a major role to play in preserving the environment.

In response to growing agricultural challenges, integrating innovative solutions into farming practices can have profound benefits for sustainable agriculture. For instance, the use of biochar in farming can significantly improve soil health by minimizing soil erosion, enhancing its water retention capacity, reducing nutrient leaching, conserving water, and boosts crop productivity (Sharma et al. 2025). Thus, adoption of such innovations into the farming practices offers a sustainable solution in building climate resilience in agriculture. However, the use of biochar remains in its early stage and require further investigation for broader implementation.

Local training for farmers is often insufficient (Goodwin and Gouldthorpe 2013). As a result, communities frequently rely solely on traditional knowledge and past experiences. This local wisdom must be recognized and integrated with modern techniques to achieve sustainable impact at both local and global levels.

To generate more local income, water sports facilities could be developed in the Kangra Valley. For example, a swimming pool has already been constructed on a khad in Upper Dharamshala. However, currently, no user fee is charged. Introducing a nominal fee could help generate funds for maintenance and further development initiatives.

Similar swimming pool like structures could be constructed along other khads. The revenue generated could be reinvested into the valley's development. Promotion of other likely water-based sports facilities would likely attract more tourists, provide another option of earning, which help in fostering regional economic growth.

Chinese philosopher Lao Tzu once said *"Give a man a fish and you feed him for a day. Teach him how to fish and you feed him for a lifetime"*. This wisdom holds true for the farmers of Kangra Valley. Rather than relying solely on external support, the emphasis should be on empowering local communities to develop and implement their own climate change solutions. By providing them with the necessary knowledge and tools, communities can develop climate-resilient livelihoods that are locally relevant and sustainable in the long term.

Furthermore, international aid is often unreliable and uncertain due to political instability, which can disrupt numerous programs related to water and food security. This unpredictability underscores the importance of prioritizing self-sufficient, sustainable local solutions through investment in community-driven initiatives. Such initiatives ensure long-term resilience and sustainability, independence from external funding fluctuations.

Furthermore, the agricultural system is increasingly hampered and compounded by the impact of climate change. For instance, rising average temperature can lead to reduce soil moisture, increased soil salinity and ultimately, a reduction in crop productivity. This situation is particularly concerning in mountainous areas where agricultural productivity is affected by climate variability. Climate extreme not only escalates food production but also continue to endanger lives and threaten communities.

In the present climate uncertainties, climate change is potentially fuelling instability and exacerbating existing challenges. It is therefore crucial for institutions to promote climate resilient crops with government subsidies. Although efforts are being made, but this should fit in, in the agro-economy to strengthen agri-businesses so that farmers receive optimal benefits from the resilient crops over traditional varieties.

Many times, climate change is often cited as the predominant cause for water and food crises. However, not all disruptions can be solely attributed to it. Human interventions such as the construction of hydropower plants and the diversion of natural water flows have significantly altered water availability. These changes often reduce downstream water supply, and ultimately limit agricultural productivity, disrupt local businesses and livelihoods. Therefore, attributing all environmental and resource-related challenges solely to climate change overlooks the human-induced modifications to natural systems.

Thus, traditional knowledge should be recognized and integrated into the planning and execution of development schemes. Although communities' involvement often gets disconnected with the institution's intervention.

Furthermore, bridging knowledge gaps, fostering trust, and promoting citizen science should be encouraged in the regions. There must be a strong focus on translating policies into practical, on-the-ground actions that deliver measurable impact. Institutions should also establish annual recognition programs to reward individuals making significant contributions to farming and rural development.

Encouraging agriculture-based startups should be encouraged. Recently, institutions are also encouraging youth of Kangra for agri-entrepreneurship,[5] which is a promising trend that should be further strengthen. Promoting such initiatives are signalling a positive shift toward innovation in agriculture sector.

Furthermore, to strengthening the livelihood, the valley must explore opportunities in other employment-generating sectors. Attracting responsible investors to the region could be a crucial step toward economic diversification and long-term sustainability.

Additionally, it is important to recognize that the Kangra Valley is located in a seismically active zone, making it highly vulnerable to earthquakes. As a result, agriculture and other natural resource-based sectors are likely at significant risk from natural disasters. This underscores the need for integrating disaster risk reduction strategies into all development and resource management plans to ensure long-term sustainability and resilience.

Under rapidly climate change conditions, it is essential for the Kangra Valley to prioritize water conservation through various feasible measures. The evident decline in rainfall over recent years underscores the urgency of this need. The valley's environmental and agricultural landscape has undergone significant changes over the past few decades, demanding a shift in approach. Therefore, our focus should be on developing a sustainable climate resilient agrarian future because climate emergencies cannot wait.

Meaningful measures, such as the collection of drain water in artificial ponds, offer a sustainable solution for water management. Additionally, constructing substantial check dams on khads is essential to recharge downstream springs and groundwater reserves. Additionally, there is a need to explore the use of modern techniques to supplement traditional water resources. For example, collection of fog water could offer a valuable resource for various uses.

The most effective way to develop effective solutions is by fostering open dialogue, raising awareness and building the capacity of local communities. Collaborative brainstorming sessions, panel discussions with institutional stakeholders and valley dwellers, will advance the discussion and can play a key role in shaping practical and sustainable policies.

---

[5] https://www.tribuneindia.com/news/himachal/kangra-youth-encouraged-to-explore-agri-entrep reneurship/ (Assessed on 01–03-2025).

5.5 Conclusion

Despite ongoing challenges, institutions are making commendable efforts to enhance agricultural productivity and preserve traditional water conservation structures in the Kangra Valley. However, to maximize the benefits of these efforts, existing schemes need to be revisited, improved, and restructured in response to current experiences and evolving conditions. Institutions must also acknowledge existing shortcomings and revise their strategies in a timely and responsive manner. The collection of real-time, low-cost data is crucial for enabling timely, informed decision-making that strengthen water governance that ensures long-term sustainability. Such efforts will not only enhance resilience to climate variability but also pave the way for a more secure and prosperous future for local communities.

This study advocates for synergistic and adaptative efforts to responsively manage the valley's natural resources. At the same time, it is to be acknowledged that formulating rigid, one-size-fits-all policies may not be feasible given the complex and dynamic nature of the region's environmental and social systems.

References

Alam MF, McClain M, Sikka A, Pande S (2024) Subsidies alone are not enough to increase adoption of agricultural water management interventions. Front Water 6. https://doi.org/10.3389/frwa.2024.1444423

Briggs J (2005) The use of indigenous knowledge in development: problems and challenges. Prog Dev Stud 5(2):99–114. https://doi.org/10.1191/1464993405PS105OA

Chinnasamy P, Maheshwari B, Dillon P, Purohit R, Dashora Y, Soni P, Dashora R (2018) Estimation of specific yield using water table fluctuations and cropped area in a hardrock aquifer system of Rajasthan, India. Agric Water Manag 202:146–155. https://doi.org/10.1016/J.AGWAT.2018.02.016

Dev R, Bali M (2019) Evaluation of groundwater quality and its suitability for drinking and agricultural use in district Kangra of Himachal Pradesh, India. J Saudi Soc Agric Sci 18(4):462–468. https://doi.org/10.1016/J.JSSAS.2018.03.002

Goodwin J, Gouldthorpe J (2013) Small farmers, big challenges: a needs assessment of Florida small-scale farmers' production challenges and training needs. J Rural Social Sci 28(1). https://egrove.olemiss.edu/jrss/vol28/iss1/3

Jamshidi O, Asadi A, Kalantari K, Azadi H, Scheffran J (2019) Vulnerability to climate change of smallholder farmers in the Hamadan province, Iran. Clim Risk Manage 23:146–159. https://doi.org/10.1016/J.CRM.2018.06.002

Marin P, Tal S, Yeres J, Ringskog K (2017) Water management in Israel: key innovations and lessons learned for water-scarce countries. www.worldbank.org/water

Rajput J, Kushwaha NL, Sikka A, Faiz Alam M, Mahapatra S, Sena DR, Singh DK, Sahoo PK, Mani I (2024) Water accounting of groundwater over exploited districts in Haryana and Punjab states to analyse impacts of water conservation measures on water availability. Water Supply 1. https://doi.org/10.2166/WS.2024.201

Sharma P, Ali S, Jayanta, Biswas K (2025) Application of biochar for soil erosion control and environmental management: implications for achieving sustainable development goals. Discover Soil 2(1):1–19. https://doi.org/10.1007/S44378-025-00065-0

Singh J (1982) Impact of irrigation on farm income and resource productivity (A study of Thural lift irrigation scheme district Kangra, H.P.) (M.Sc Thesis, Himachal Pradesh Krishi Vishva Vidyalaya).

Thakur D, Bartarya SK, Nainwal HC (2018) Tracing ionic sources and geochemical evolution of groundwater in the Intermountain Una basin in outer NW Himalaya, Himachal Pradesh, India. Environ Earth Sci 77(20):1–23. https://doi.org/10.1007/S12665-018-7891-7

Thakur SD, Panda AK (2012) Laboratory investigation of drinking water sources of Kangra, Himachal Pradesh. J Commun Dis 44(2):103–108. https://europepmc.org/article/med/25151755

Questionnaires

S. No.	Question[1]	Answer
1	Agriculture and water as matters of polling during elections	100% respondents said yes
2	Reliance on government schemes	50% said yes[2]
3	Water collection tanks	Presence is negligible
4	Landholding in the Kangra Valley	Majority of farmers have small or limited landholdings
5	Water availability in streams 20 years ago	100% agreed it was better in the past
6	Average annual income from 1 acre of land	Ranges between ₹85,000–₹1,50,000[3]
7	Factors leading to agricultural water crises	100% agreed on weak and fragmented policies
8	Declining rainfall pattern	100% agreed rainfall has declined
9	Increase in average temperature	100% said yes
10	Increase in frequency of climate events	100% agreed frequency has increased
11	Is micro-irrigation a real solution?	90% were unsure
12	Are farmers continuing agriculture or shifting occupations?	50% said they are shifting

(continued)

---

[1] The questions were asked from a limited sample size and may not fully represent the views of all valley dwellers.

[2] Some valley dwellers preferred to avoid this question.

[3] 1 US dollar is equal to ₹85.5 rupees (as on 05–04–2025).

S. Ali, *Traditional Water Conservation Community-Managed Structures and Their Role in Valley Dwellers' Livelihoods*, SpringerBriefs in Water Science and Technology, https://doi.org/10.1007/978-3-032-04637-6

(continued)

S. No.	Question	Answer
13	Farmers' training on water management practices	Only 10% have received training[4]
14	Awareness of how to respond to climate change	Only 10% said yes
15	Awareness of irrigation schemes and subsidies	50% are aware
16	Willingness to pay for water-resilient crops at subsidized prices	Only 25% are willing
17	Present condition of agriculture	100% agreed agriculture has substantially reduced in the valley

---

[4] Farmers expressed discomfort in answering this question. They do not want to get in trouble by highlighting any shortcomings of institutions.

If you have any concerns about our products,
you can contact us on
ProductSafety@springernature.com

In case Publisher is established outside the EU,
the EU authorized representative is:
Springer Nature Customer Service Center GmbH
Europaplatz 3, 69115 Heidelberg, Germany

Printed by Libri Plureos GmbH
in Hamburg, Germany